T0387121

BUTTERFLIES
OF THE WORLD

BUTTERFLIES
OF THE WORLD

A GUIDE TO EVERY SUBFAMILY

Blanca Huertas
Shinichi Nakahara

PRINCETON UNIVERSITY PRESS
PRINCETON AND OXFORD

Published in 2025 by Princeton University Press
41 William Street, Princeton, New Jersey 08540
99 Banbury Road, Oxford OX2 6JX
press.princeton.edu

Conceived, designed, and produced by
The Bright Press
an imprint of The Quarto Group
1 Triptych Place, London, SE1 9SH, United Kingdom
www.Quarto.com

GPSR Authorized Representative: Easy Access System Europe – Mustamäe tee 50,
10621 Tallinn, Estonia, gpsr.requests@easproject.com

Library of Congress Control Number: 2024949045
ISBN: 978-0-691-26520-9
Ebook ISBN: 978-0-691-26710-4
British Library Cataloging-in-Publication Data is available

Publisher **James Evans**
Editorial Director **Isheeta Mustafi**
Art Director and Cover Design **James Lawrence**
Managing Editor **Lucy Tipton**
Senior Editor **Joanna Bentley**
Project Editor **David Price-Goodfellow**
Design **Ginny Zeal**
Picture Research **Susannah Jayes**
Illustrations **John Woodcock**
Production Controller **George Li**

Cover photos: front, clockwise from top left: Shutterstock/Philippe Clement, /James
Laurie, © Copyright The Trustees of the Natural History Museum, London, Jean-Yves
Gallard, Shutterstock/Billion Photos, Dreamstime/Feathercollector, Shutterstock/
Mirek Kijewski, Alamy/Papilio, Shutterstock/Pyty, /Feathercollector, /M Shcherbyna,
/Super Prin. Back and spine: Shutterstock/Vitalii Hulai, /Billion Photos.

Printed in Malaysia

10 9 8 7 6 5 4 3 2 1

INTRODUCTION

Butterflies are undoubtedly charismatic insects; their aesthetic beauty has stimulated a wealth of enthusiasm for centuries, cementing their reputation for being the best-studied group of insects. Unlike many other insects, butterflies appeal greatly to the public due to their aesthetic value and conjure a positive image. People of all ages enjoy seeing butterflies flying in their own yard or nearby park, and as children a few of us may have enjoyed raising a butterfly from an early stage and releasing the adult. Butterflies have even starred in movies. Most butterflies are herbivores as caterpillars, and as pollinators, the adults have an important role in ecosystems. These multidimensional features help to set them apart from other insect groups.

One way to introduce someone to the delights of butterflies is to share those breathtaking moments of observing a spectacular butterfly in nature. The nymphalid genus *Agrias* (classified as *Prepona* by some scientists), found in the rainforests of Central and South America, is a group of brightly colored butterflies that exhibit a great degree of individual variation. In fact, the variation is so diverse that each specimen appears to be different from one another. Because of this variation, hundreds of names have been proposed for this genus even though only a few species are recognized. Their beauty and variety, coupled with their rarity, have captured the hearts of naturalists, collectors, and scientists for centuries.

A tribute to their beauty was summarized by German entomologist Hans Fruhstorfer, who started his career by collecting in Brazil in the late 1800s and subsequently wrote: "In this magnificent tropical genus [*Agrias*], upon which nature seems to have showered all her abundance of most brilliant colours … is undoubtedly one of the most magnificent sights that nature has ever produced in the whole world of butterflies." Collectors such as Fruhstorfer traveled extensively, examining and describing countless exotic tropical butterflies.

Butterflies have a remarkable power to inspire people with a love of nature, and we hope that this book will spark a comparable interest in butterflies in those who are just beginning their discovery of these marvelous insects.

OPPOSITE | The Old World Swallowtail (or simply "Swallowtail") *Papilio machaon* is a conspicuous and widespread butterfly. Despite its common name, it is also found in North America.

BELOW | White-spotted Agrias *Agrias amydon* feeding on an unknown substance on a leaf. *Agrias* spend most of their time in the canopy so observing them typically requires attracting them with bait.

WHAT ARE BUTTERFLIES?

On these expanded membranes [butterfly wings] Nature writes, as on a tablet, the story of the modifications of species, so truly do all changes of the organisation register themselves thereon. Moreover, the same colour-patterns of the wings generally show, with great regularity, the degrees of blood-relationship of the species. As the laws of nature must be the same for all beings, the conclusions furnished by this group of insects must be applicable to the whole world.

From: Henry Walter Bates,
The Naturalist on the River Amazons

Butterflies (and moths) are insects classified in the order Lepidoptera, and approximately 20,000 species of butterflies are known worldwide, though none are found in Antarctica. Moths are about seven times as diverse as butterflies, but butterflies are often more conspicuous and are associated with sunny days, prompting people's fascination and enjoyment as they watch them visit flowers in their yard or local parks.

Butterflies are studied extensively. They include taxa with biological significance, such as the *Heliconius* butterflies, *Papilio dardanus*, and the Monarch butterflies *Danaus plexippus* (more on these later). The excerpt on the left was written by Henry Walter Bates, a Victorian naturalist; it is taken from his book *The Naturalist on the River Amazons*, based on his observations in the Brazilian Amazon.

Bates discovered an explanation for the diversity of life by observing that two distantly related butterfly species had very similar wing patterns. In particular, Bates observed that aposematic (with warning coloration) butterflies were able to avoid predation in the rainforest, and many other nontoxic butterflies and moths that occur in the same areas appear to imitate butterflies that use chemical defense mechanisms. This phenomenon is known today as Batesian mimicry in his honor. It is an adaptation in which palatable species take on features of distasteful ones in order to fool predators and gain protection. Batesian mimicry is a striking example of evolution by natural selection, a theory formed around the same time, independently, by English naturalists Charles Darwin and Alfred Russel Wallace.

Bates continued: "The study of butterflies—creatures selected as the types of airiness and frivolity—instead of being despised, will someday be valued as one of the most important branches of biological science."

More than 150 years have passed since then, and indeed butterflies have been an important study group in numerous ecological, evolutionary, and biogeographical studies. However, the value and impact possessed by butterflies reach far beyond the scientific community. Many of us read Eric Carle's *The Very Hungry Caterpillar* as children, and learned the life-stages of a butterfly as it passes through egg, caterpillar, and pupa until it reaches the adult stage. The book has been translated into more than 70 languages and read worldwide. Butterflies conjure positive images of peace and a healthy environment. We can find many references to butterflies throughout history

ABOVE | *Heliconius* butterflies and many other toxic butterflies are warningly colored, visually signaling their distastefulness to predators.

in various nonscientific contexts, such as butterfly-shaped accessories from ancient Mesopotamia to modern jewelry, poems from early eleventh-century Heian-period Japan, engravings in Indigenous artifacts depicting butterfly wings, and even in the name of a swimming stroke!

THE ORDER LEPIDOPTERA

LEFT | The *Urania* swallowtail moth is diurnal and can be spotted puddling or attracted to bait with butterflies in the Amazon rainforest.

The order Lepidoptera includes butterflies and moths. The name *Lepidoptera* is derived from two Greek words: *lepido* (scales) and *pteron* (wings), in reference to their scaly wings. It is closely related to the order Trichoptera, commonly known as caddisflies. Unlike Lepidoptera larvae, caddisfly larvae are aquatic, and the wings of Trichoptera adults are covered with hairs, not scales, hence they are called Trichoptera: from *trichon* (hairs). Lepidoptera and Trichoptera have been shown by molecular studies to be closely related to Diptera (flies and mosquitoes), Siphonaptera (fleas), and Mecoptera (scorpion flies and hanging flies). These orders all belong to a group of insects characterized by the immature stages being distinct from the adults. They undergo four life-stages: egg, caterpillar (larval stage), chrysalis (pupal stage), and adult (usually winged) butterfly. The process of passing through these four distinct life-stages is scientifically termed *complete metamorphosis*.

Historically, as well as scientifically, seven families have been accepted as belonging to the true butterflies, known as the superfamily Papilionoidea: Papilionidae (including swallowtails, Apollo, and festoons), Pieridae (whites and yellows), Lycaenidae (blues, hairstreaks, and coppers), Riodinidae (metalmarks), and

LEFT | Unlike Lepidoptera, caddisfly (Order Trichoptera) larvae are aquatic. Adult caddisflies (shown here) can be found attracted to lights.

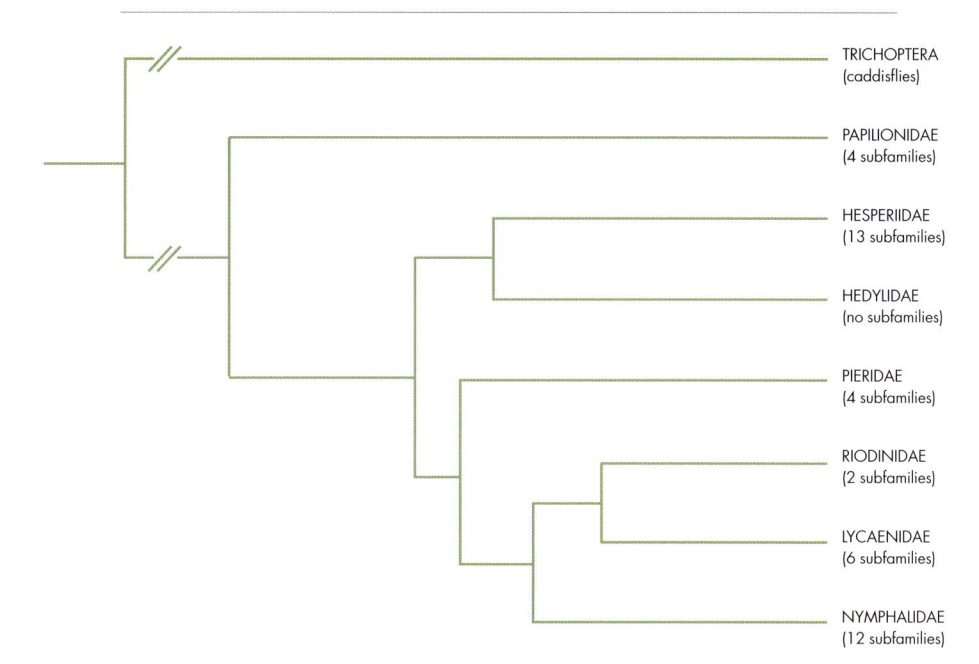

TRICHOPTERA
(caddisflies)

PAPILIONIDAE
(4 subfamilies)

HESPERIIDAE
(13 subfamilies)

HEDYLIDAE
(no subfamilies)

PIERIDAE
(4 subfamilies)

RIODINIDAE
(2 subfamilies)

LYCAENIDAE
(6 subfamilies)

NYMPHALIDAE
(12 subfamilies)

Nymphalidae (brush-foots), and (included more recently) Hesperiidae (skippers) and Hedylidae (moth-like butterflies). In appearance, skippers are somewhere between butterflies and moths. Their classification as butterflies is well supported, largely by DNA analysis. The moth-like butterflies (Hedylidae), however, look just like typical drab moths and were classified as such until 1986. Recent molecular data corroborates this classification, which may seem counterintuitive.

How can butterflies and moths be distinguished from each other? This is a question that scientists have a very hard time answering. A stereotypical view might be that butterflies are "pretty, winged insects flying during the day" and moths are "drab things flying at night." There are other characteristics proposed to separate the two groups, such as the shape of the antennae and wing coupling types, but there are always exceptions.

According to recent phylogenetic studies, butterflies are day-flying moths, or vice versa,

moths are night-flying butterflies. Indeed, there are languages that do not really distinguish between butterflies and moths: in German (biologically) they are all *schmetterlinge* and in Indonesian *kupu-kupu*. The colorful day-flying swallowtail moths in *Urania* were originally classified as butterflies (*Papilio*) by Carl Linnaeus, the father of taxonomy, in his classic book *Systema Naturae*. So, indeed, the boundary appears to be blurry. However, the separation between the two seems to be generally accepted, and different words for them coexist as mutually exclusive terms in many languages, including English, indicating that we recognize and communicate about these two groups as different entities; and we will continue to do so, despite science's counterintuitive classification. Science is not the only valid way of classification, but using scientific names and separating them into groups makes it easier to communicate about organisms and their unique characteristics.

EVOLUTION AND ORIGINS

In introductory biology textbooks, a typical explanation of evolution is that it is a bifurcating process, with species arising from a common ancestor in tree-like patterns. However, a very recent study on three *Heliconius* butterflies called this conventional depiction of the evolutionary process into question by suggesting that new species can evolve through merging of lineages. In other words, through hybrid speciation. Researchers showed that the *Heliconius elevatus* genome is a mixture of two extant parental species, *H. melpomene* and *H. pardalinus*, and that the hybridization took place around 180,000 years ago. The "tree" of life does not in fact resemble a tree but appears to be a reticulation network, a notion supported mostly by examples in plants.

At a much larger scale, researchers have been investigating how butterflies got to where they are today. Recent study suggests butterflies evolved ~100 mya and diverged from a group of nocturnal plant-feeding moths in the mid-Cretaceous. Since then, butterflies have diversified and we see butterflies on all continents today except for Antarctica; they can be found in various habitats ranging from a small park in an urban area to snow-capped mountains. Like moths, having wings and the ability to go through complete metamorphosis (holometabolous development) contributed greatly to their evolutionary success. They can exploit different niches, such as shifting host plants (plants where eggs are laid and/or caterpillars feed) at different life-stages and can be flexible during their immature stages by delaying growth if environmental conditions are not suitable. When they reach the adult stage, they develop wings and some species have the ability to migrate and disperse. Butterflies' colorful wing patterns also contributed toward their evolutionary success, allowing mate recognition, signaling, and predator avoidance (e.g., camouflage, mimicry).

Research on butterflies has shown us how organisms can change through time and space.

LEFT | An imaginary example of "cladogenesis". Cladogenesis describes the process of speciation by lineage splitting (i.e., a bifurcating evolutionary process).

OPPOSITE BOTTOM | Snow-capped peaks of the Andes in Colombia. The giant rosette plants characterize the "Páramo," a high mountain grassland found in the northern part of the tropical Andes. *Lymanopoda samius* (**inset**) is one of many characteristic butterflies found in this area.

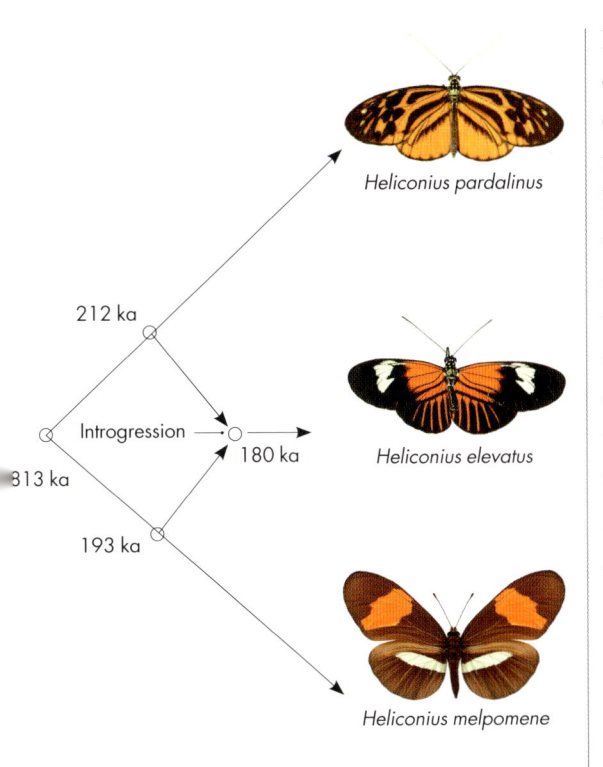

Heliconius pardalinus

212 ka

Introgression → 180 ka

Heliconius elevatus

313 ka

193 ka

Heliconius melpomene

Like many other organisms, butterflies are not evenly distributed around the globe. Some parts of the world harbor more species of butterflies than others. We see the highest diversity of butterflies in the Neotropical region (Central and South America, and the Caribbean), where approximately 40 percent (8,000 species) of the world's butterfly species are concentrated. The disproportionate butterfly species richness observed in the Neotropics has prompted people to study this region to determine why it harbors so much of Earth's biodiversity. Molecular analysis has led to an emerging consensus that the tropical Andes contributed greatly toward speciation events by acting as a "species pump," resulting in the diversity of the Neotropics. But not all butterfly diversifications can be explained by the uplift of the Andes.

ABOVE LEFT | *Heliconius elevatus* arising from its two parental species (*H. pardalinus* and *H. melpomene*) based on a hybridization that occurred approximately 180,000 years ago (180 ka).

ANATOMY OF ADULTS AND COLOR/VISION

The body of an adult butterfly, as in other insects, has three main parts: head, thorax, and abdomen. The main components of the head are the compound eyes, antennae, labial palpi, and the proboscis (tongue). The thorax is divided into three segments: prothorax, mesothorax, and metathorax. These segments are each accompanied by a pair of legs: prothoracic leg (foreleg), mesothoracic leg (mid leg), and metathoracic leg (hind leg). Each of these legs is composed of coxa, trochanter, femur, tibia, and tarsus. Forewings and hindwings are articulated to the meso- and metathorax respectively. The abdomen is attached to the metathorax and is made up of ten segments, the ninth and tenth being the genitalia.

SCALES

Butterflies (and moths) are characterized by having scales on their wings and other parts of the body. Their scales contribute significantly to the display of colorful, metallic, or iridescent markings in many species, with some specialized as secondary sexual characters. Adult butterfly scales are cuticular in structure; each has a short stalk extending from an individual socket and expands into a blade-like shape. Scales are laid out neatly on the wings with some overlap and variation in size and shape, but the dorsal (upper) lamina is often accompanied with an intricate network of longitudinal ridges and cross ribs. The ventral (under) lamina is typically smooth. Scales of adult butterflies are often hollow to some extent. These

EXTERNAL ANATOMY OF A BUTTERFLY

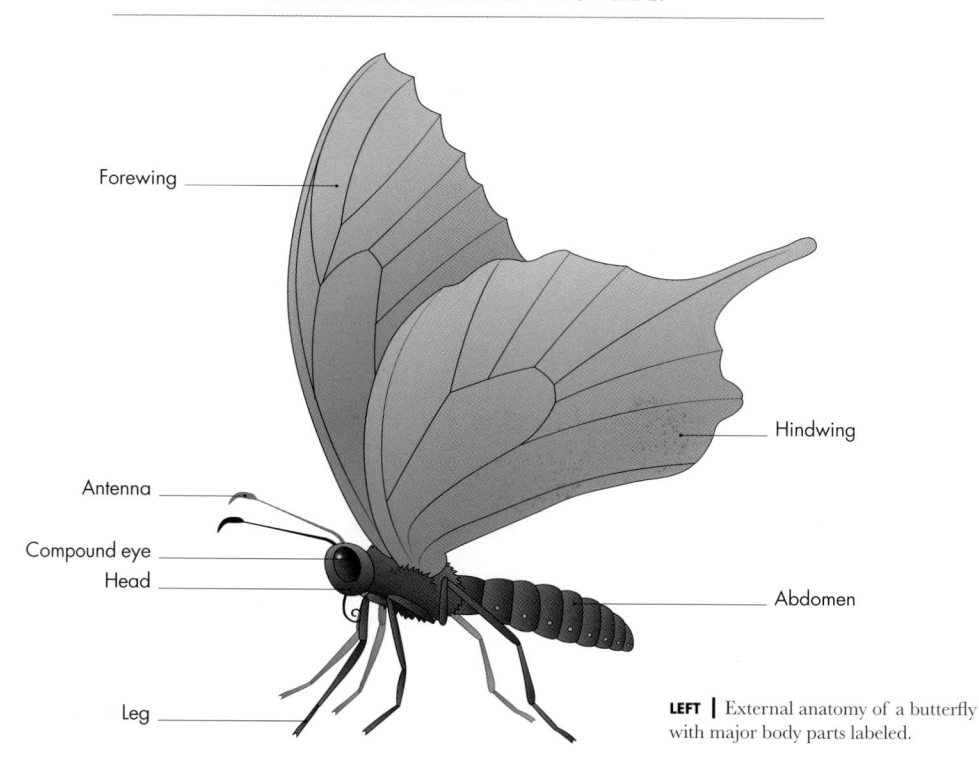

Forewing

Hindwing

Antenna

Compound eye

Head

Abdomen

Leg

LEFT | External anatomy of a butterfly with major body parts labeled.

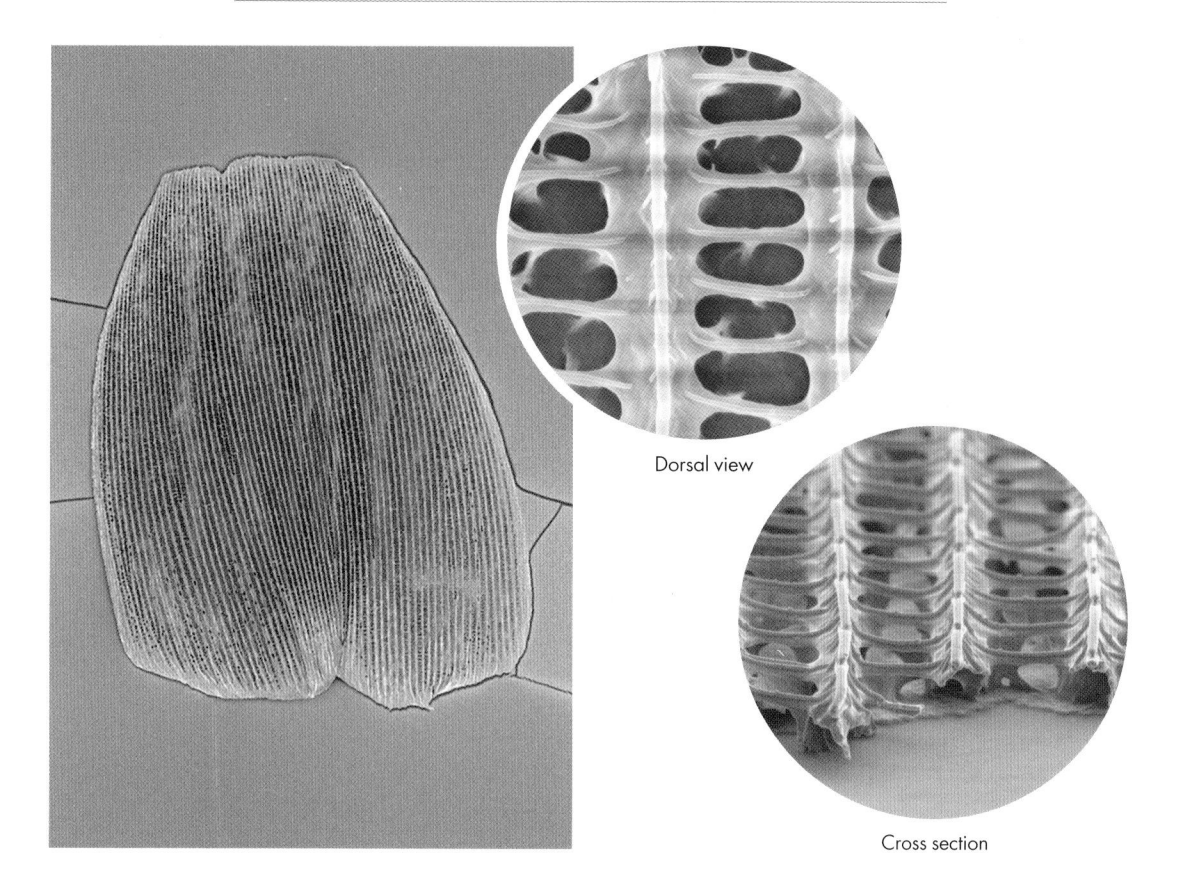

Dorsal view

Cross section

scales can be classified into many different types based on their function and structure. Each scale is responsible for producing a single color, which can be generated by pigment deposition and/or the scale's structure. The conspicuous wings of blue morphos (Nymphalidae) are particularly beautiful. Their blue-lilac colors are generated by minute physical structures in the scales. Some adult male butterflies have specialized scales called *androconia*, involved in the dissemination of scents or pheromones to lure females. These androconial scales are often found on the wings, but they can be located on the abdomen or even in the genitalia, and they come in different shapes and forms: some are patch-like, feather-like, or like a crest or a tuft (termed a *hair pencil*).

ABOVE | Scanning electron micrograph (SEM) of a typical butterfly wing scale. Close-up view of the dorsal side and cross-sectional view.

WING VENATION

Butterfly wings are membranous and composed of two layers, in addition to thickened tubular structures called veins running through the membranes and providing support. The veins are filled with fluid from the hemolymph and are thought to play an important role during expansion of wings after eclosing from the pupa. Recent studies suggest that an inflated portion of the wing vein seen in wood-nymphs (Satyrinae) has a hearing function.

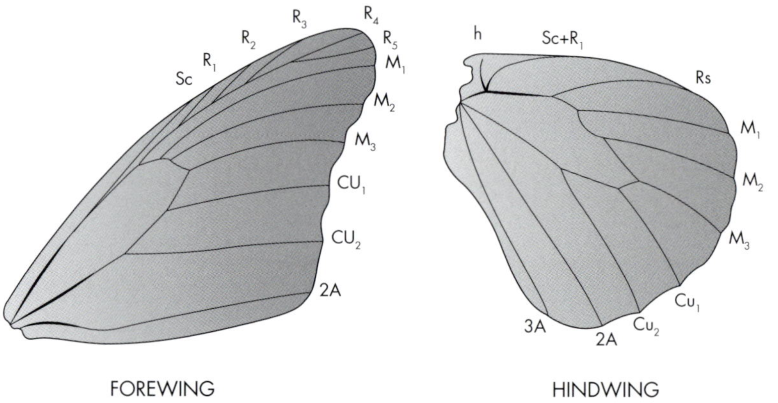

BUTTERFLY WING VENATION

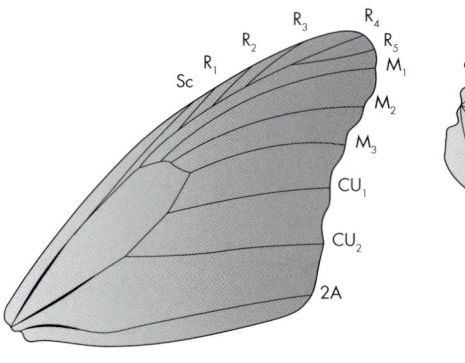

FOREWING

Labels on forewing: Sc, R₁, R₂, R₃, R₄, R₅, M₁, M₂, M₃, CU₁, CU₂, 2A

HINDWING

Labels on hindwing: h, Sc+R₁, Rs, M₁, M₂, M₃, Cu₁, Cu₂, 2A, 3A

ABOVE | Butterfly wing venation with veins conventionally labeled.

BELOW | Close-up view of butterfly wing veins covered with tiny scales.

More than a hundred years ago, a few entomologists noticed that the wing venation of insects is useful in higher-level classification and tried to come up with uniform terminology for them. This resulted in a numbering scheme called the Comstock–Needham system, named after entomologists John Henry Comstock and George Needham. This system can also be used to diagnose and identify butterfly families.

The family Papilionidae (swallowtails) have a short second anal vein on the forewing, running from the base of the wing toward the inner margin. Other butterfly families have only one forewing anal vein, except for the skippers. Swallowtails also have a short cross-vein between the cubitus (Cu) and the first anal vein, but this vein is not present in the apollos (Parnassiinae). The family Pieridae can be recognized by the forewing vein M_1 not originating from the discal cell, but instead merged into the radial veins (R_{2-5}). Lycaenidae and Riodinidae typically have fewer than five radial veins on the forewing. The two radial veins R_1 and R_2 of Hedylidae are sinuous and appear as if they might touch each other. Occasionally, wing venation can even be informative at species-level classification of butterflies. A recently described new genus of a South American butterfly, *Xenovena* (Satyrinae) is named for the distinctive hindwing venation of its only species, *X. murrayae*. *Xenos* means "strange" in Greek and *vena* means "vein" in Latin.

ANTENNAE

A classic diagnostic feature for distinguishing butterflies (other than hesperids and hedylids) from moths is their antennae. The majority of

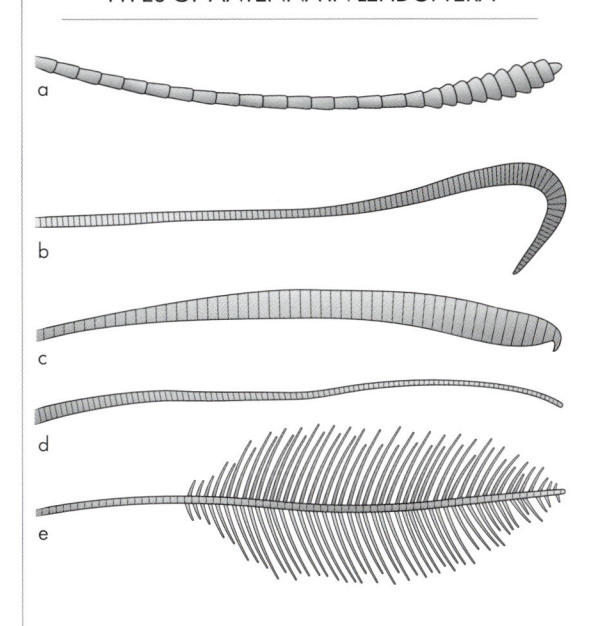

ABOVE | Variation of antennal structure within Lepidoptera.
a. Butterfly
b. Skipper butterfly
c–e. Moths

butterflies have clubbed or knobbed antennae, whereas antennae shape in moths is quite diverse: they may be feathery, wispy, thread-like, or spindly. It is thought that the rather uniform antennae of butterflies compared with more

LEFT | Close-up of the hooked portion of a Hesperiidae antenna showing the part without scales (termed the nudum).

diverse and complex moth antennae is a result of moths' reliance on scent rather than vision, because most are nocturnal. But there are always exceptions, and not all butterflies and moths can be separated by antennae inspection. Many skipper butterflies (Hesperiidae) have relatively short and hooked antennae, and the tips typically lack scales. Adult skippers also tend to be drab, with a moth-like, robust (chubby) body. On the other hand, moths of the family Castniidae are known to be diurnal or crepuscular, and some species are brightly colored and have clubbed antennae, just like butterflies.

EYES AND VISION

Butterflies do not rely heavily on vision when they are in the caterpillar (larval) stage. Caterpillars have six simple eyes (stemmata) on each side of the head, which they use to receive light, but their vision at the larval stage is limited. Adult butterflies have a pair of compound eyes, each of which is a cluster of numerous units called ommatidia. These ommatidia include multiple photoreceptor cells. Like humans and unlike caterpillars, adult butterflies rely heavily on color vision for certain vital activities, such as flower foraging. However, unlike the red-green-blue (RGB) trichromatic retinas of humans, butterflies are known to possess multiple photoreceptors. In the same way that not all humans perceive the world alike, butterflies also see the world differently.

IDENTIFYING FEATURES

Some external morphological characters of adult butterflies can be used to diagnose families and subfamilies. The labial palpi of snouts (Libytheinae) are developed and protruding, resembling a long nose (snout). The foreleg of brush-foots (Nymphalidae) is greatly reduced and useless for walking; they thus appear to have only two pairs of legs: mid leg and hind leg. The foreleg of males of some metalmarks (Riodinidae) is also reduced (normal in females), and this is one feature that can be used to separate the metalmarks from the blues (Lycaenidae). These two families were historically considered grouped together in the family Lycaenidae.

GENITALIA

Taxonomists have used the genitalia of males as a way to help identification and separate similar species morphologically. These reproductive organs often show species-specific characteristics. The study of male butterfly genitalia became increasingly common in the mid-twentieth century, and it is standard to illustrate the genitalia in descriptions of new species. Study of female genitalia has advanced more recently, but is still undertaken less often than in males. Females of many tropical species remain undiscovered.

RIGHT | A male genitalia illustration prepared by one of the authors (SN). While photographing genitalia is common today, drawing the genitalia using camera lucida is still done by some taxonomists.

LIFE-STAGES

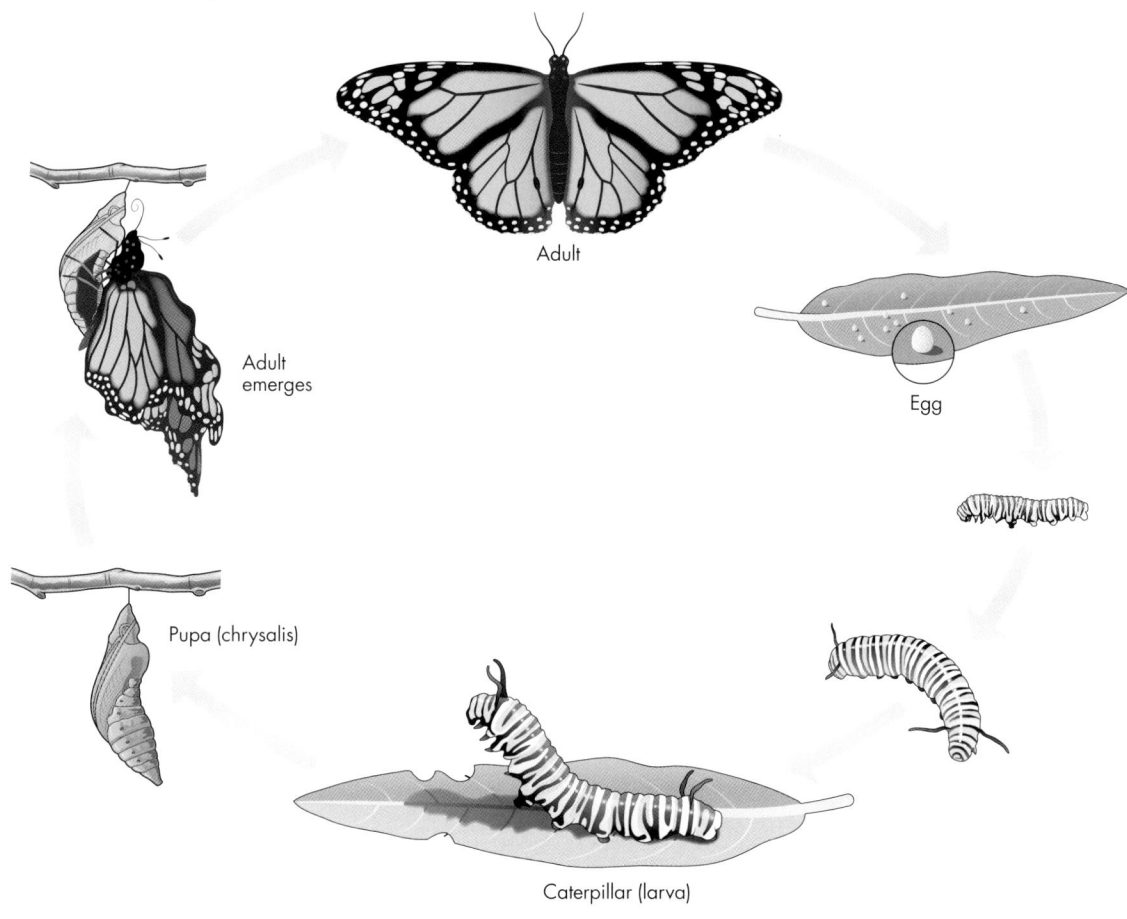

Adult

Adult
emerges

Egg

Pupa (chrysalis)

Caterpillar (larva)

ABOVE | Lifecycle of the Monarch
Danaus plexippus, typical of all butterflies.

More than 300 years ago, Maria Sibylla Merian studied immature stages of butterflies and moths and observed that they develop from an egg, through caterpillar and pupa, to reach the adult stage. At a time when people believed in spontaneous generation (i.e., caterpillar and pupa were thought to be two different organisms), her documentation provided the evidence for their complete metamorphosis. The lifecycle of butterflies, going through four stages (egg, caterpillar, pupa, and adult), is now common knowledge, but there was a time when people

were unaware of it. Since Merian's documentation, our knowledge of butterfly life-stages has improved but we still do not know what many species of butterflies feed on in the wild, especially in the tropics.

EGG

Butterfly eggs are laid singly or in batches on leaves, branches, buds, or trunks of the host plant, which can be herbaceous or woody. Many butterflies seem to have preferences for how and where they lay their eggs. The hairstreak

TYPICAL BUTTERFLY EGG

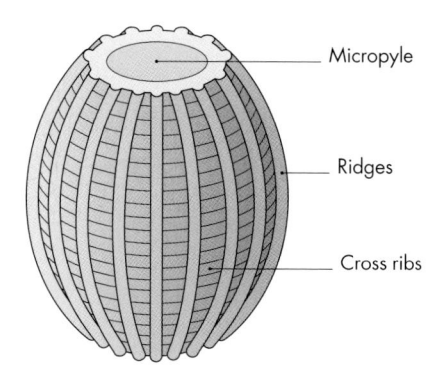

- Micropyle
- Ridges
- Cross ribs

VARIATION OF EGG STRUCTURE WITHIN BUTTERFLIES

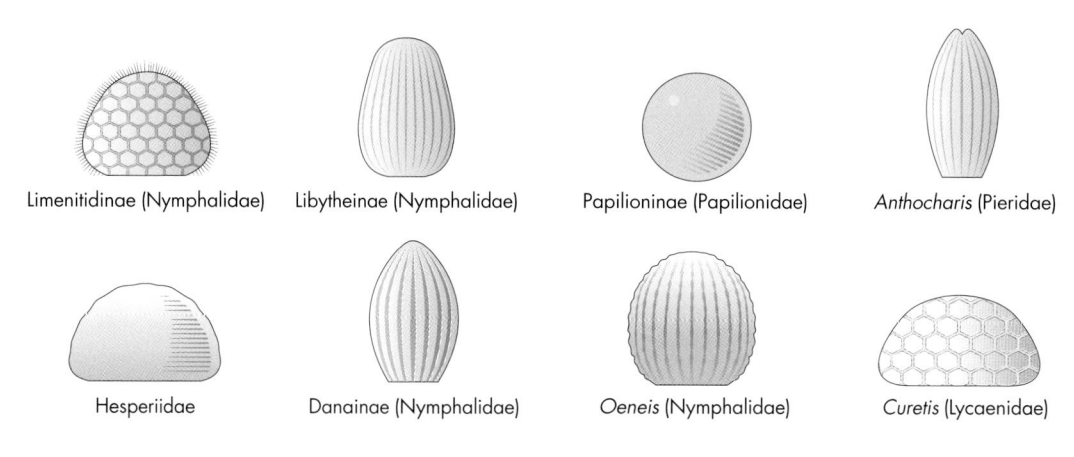

Limenitidinae (Nymphalidae)

Libytheinae (Nymphalidae)

Papilioninae (Papilionidae)

Anthocharis (Pieridae)

Hesperiidae

Danainae (Nymphalidae)

Oeneis (Nymphalidae)

Curetis (Lycaenidae)

ABOVE | Lateral views of several butterfly eggs to show the variation in shape.

Ussuriana stygiana (Lycaenidae) deposits its eggs in clusters toward the base on the trunk of Chinese/Japanese Flowering Ash *Fraxinus sieboldiana* (Oleaceae) trees. The Wonderful Hairstreak, *Thermozephyrus ataxus* (Lycaenidae), lays its eggs by the buds of the Japanese Evergreen Oak *Quercus acuta* (Fagaceae). These hairstreaks spend most of their life as eggs. The eggs are laid by females during the summer, and they overwinter in that stage until the following spring when the egg synchronizes hatching with the budding of the tree. In the tropics, on the other hand, eggs can hatch a couple of days after oviposition. Butterfly eggs are usually no greater than a few millimeters in diameter, with their shape ranging from spherical to hemispherical, spindle-like (common in the Pieridae), or oval. The chorion (outer shell) may be decorated with an intricate system of ridges and ribs as well as spines, and these decorations can be species specific. Studies assessing fecundity of a few pierids based on egg numbers indicated that several hundred eggs are present inside the abdomen in freshly eclosed females.

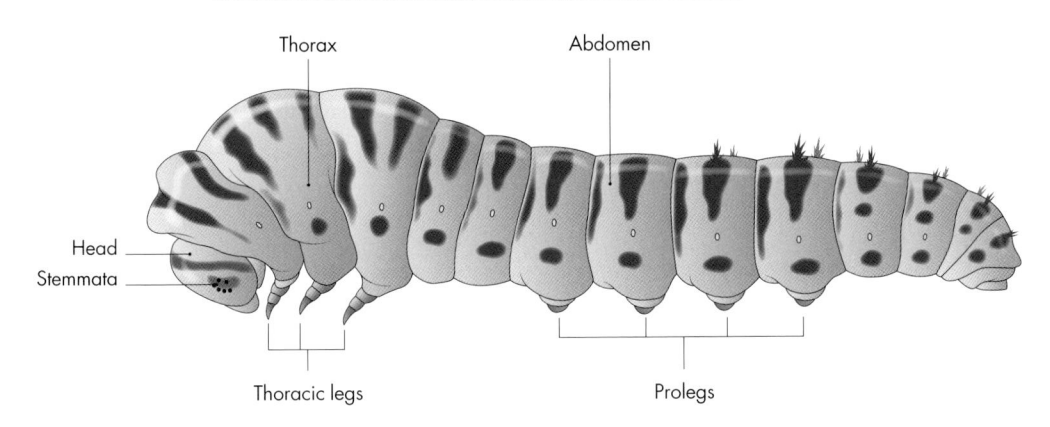

VARIATION OF CATERPILLARS WITHIN BUTTERFLIES

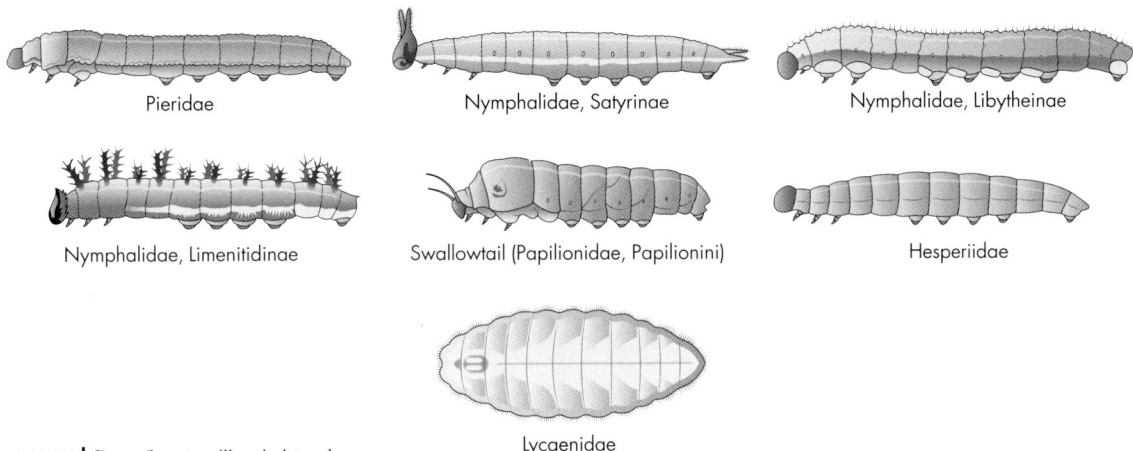

ABOVE | Butterfly caterpillars in lateral view, with the exception of the lycaenid caterpillar, which is viewed from above.

CATERPILLAR

Butterfly larvae are commonly known as caterpillars. They are usually herbivorous and are often involved in a close evolutionary relationship with the plants they feed on, resulting in some degree of selectivity. Caterpillars of Monarchs are picky and feed exclusively on milkweed plants, genus *Asclepias* (Apocynaceae), whereas the caterpillars of some skippers in the Neotropics can feed on a few different plant families. Butterflies whose caterpillars feed on a wide variety of plant families are termed polyphagous,

and those whose host plants are from a specific genus are known as monophagous. But not all butterfly caterpillars feed on plants: some are carnivorous. The Harvester *Feniseca tarquinius* (Lycaenidae) feeds on aphids throughout its larval stage; it is the only known carnivorous butterfly caterpillar in North America. In Asia, the Forest Pierrot *Taraka hamada* (Lycaenidae) also feeds on aphids as a larva, and the adults are also known to suck secretion from these aphids. Interestingly, there are butterflies that are halfway between herbivores and carnivores. These butterflies, concentrated in two families, Lycaenidae and Riodinidae, are *myrmecophilous* ("ant loving") and have intricate relationships with ants during their larval stages. Caterpillars of these butterflies have specialized organs that secrete chemical compounds to fool ants. Some species are fed by ants, some are protected by them, and some of them even live inside ant nests and eat them. Another unique larval biology can be seen in the skippers. Unlike other butterflies, skipper caterpillars spend much of their time in shelters constructed by cutting, rolling, folding, and tying host plant leaves. They even shoot excrement outside while they're inside these shelters.

Butterfly caterpillars typically pass through four to six developmental stages, known as *instars*. Many are solitary but some are gregarious. Gregarious behavior in butterflies is often associated with aposematism. Gregariousness of caterpillars varies; some are gregarious during early instars and become solitary as they develop. The duration of the larval stage is usually several weeks but it can last several months or more than a year. The Asamana Arctic or Norse Grayling *Oeneis norna* (Satyrinae) lives in a harsh environment in the Japanese Alps and spends two winters as a larva.

PUPA

The word *pupa* comes from a Latin term for "doll," referring to the transformation stage prior to the adult stage in butterflies. The term *chrysalis* is used interchangeably with *pupa* in most contexts, but *chrysalis* is applicable only to butterflies (not to moths). This stage allows the drastic transformation from caterpillar to adult, two forms that are different enough to make people think these are two different organisms. Inside the chrysalis, caterpillar tissues are degraded (some organs remain intact) by enzymes

ANATOMY OF A BUTTERFLY PUPA

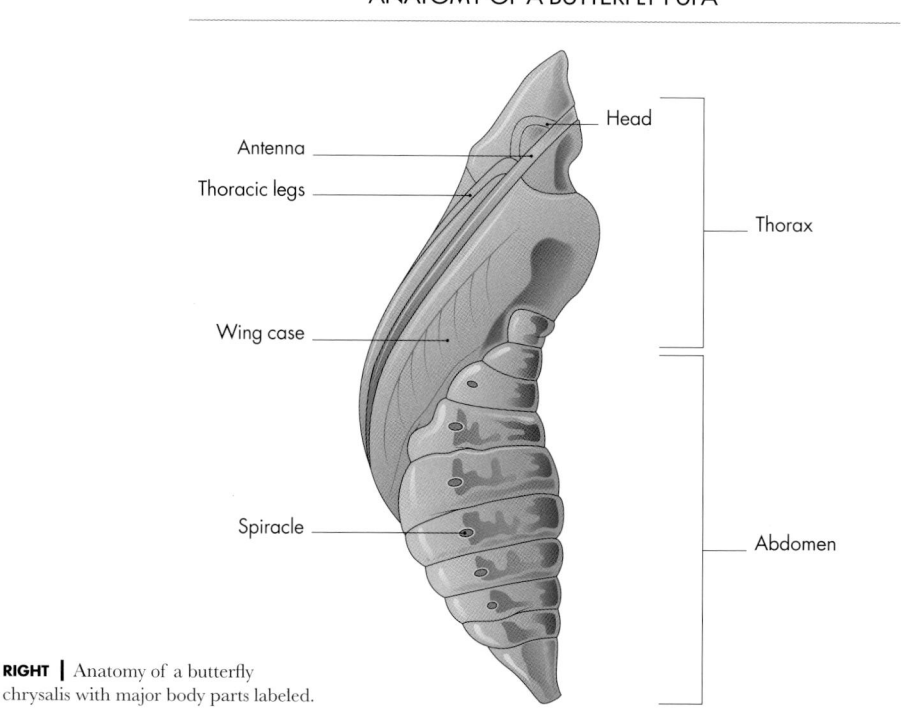

Antenna

Thoracic legs

Head

Thorax

Wing case

Spiracle

Abdomen

RIGHT | Anatomy of a butterfly chrysalis with major body parts labeled.

and produce "imaginal disks," which turn into various parts of the adult butterfly body. During this process, the chrysalis is immobile and defenseless. Many other holometabolous insects have their own protection strategy for the pupa (e.g., a moth's cocoon). However, a butterfly's chrysalis lacks this kind of protection, explaining why it tends to be camouflaged with its environment. Butterfly chrysalises are anchored at the terminal end of the abdomen with a structure called the *cremaster*, which hooks the silk pad produced by the caterpillar onto its support structure (e.g., a twig). Some species additionally produce a girdle (silk thread) to wrap around the thorax in order to support the chrysalis. Like other life-stages, the duration of butterfly pupal stage varies depending on the species, environmental conditions, and geographical location, and can also vary between broods.

ABOVE | Comparison of the types of attachment of butterfly pupation: (left) a nymphalid pupa attached to a leaf with a cremaster; (right) a swallowtail pupa supported by a silk thread.

ADULT

The goal during the adult stage (known as *imago*) is to mate and pass on the genes to the next generation. Behaviors exhibited by adult butterflies are due to their efforts to increase the probability of reproductive success. Adult males of many butterflies are known to engage in *puddling*, where a freshly eclosed male (or a group of males) swarms around mud or puddles and sips water. While puddling is widely known, it is not well understood why butterflies perform such a behavior. One long-standing hypothesis is that they do it to replenish levels of sodium chloride, which is considered crucial for mating

ABOVE | Kite-Swallowtail butterfly *Eurytides* puddling in the Amazon region of Ecuador.

LEFT | A male *Heliconius* butterfly sitting on a pupa of a female, waiting for her emergence to mate.

success; this explains why young males (which are likely to be looking for mates) perform puddling behavior far more commonly than worn ones (which may have already mated) or females. Male butterflies need a lot of energy in order to go searching for females (and look good while they do so!). Humans' sweaty clothes can attract desperate butterflies as well, perhaps for the same reason. What do they do if there are no places to puddle? Some skippers and nymphalids are known to rehydrate dry, solid substances such as bird droppings to dissolve them and sip from the fluid they create.

Once they are ready to seek females, male butterflies use various, often species-specific strategies to efficiently find partners. Unlike moths, whose females produce pheromones to lure males, male butterflies have to work hard to find a mate. Some nymphalids and lycaenids display territorial behavior whereby a male perches on a leaf in his territory and chases intruders away from it until a female comes along. Another rather common strategy is patrolling, where male butterflies fly ceaselessly in search of females. Some swallowtails have a well-defined path that they follow when engaged in patrolling, whereas satyrine species seem to just wander around the forest. When a female has been located, the male starts a sequence of courtship rituals to draw her attention, typically by flapping around the female or releasing pheromones. If the proposal is accepted, the female will close her wings and protrude her abdomen for coupling. Just like humans, finding "the one" is also not easy for butterflies. Males of *Heliconius* butterflies are known to swarm around a female pupa, waiting for her to emerge from the chrysalis!

ABOVE | A pair of Monarch butterflies
Danaus plexippus mating. The courtship
behavior typically follows a sequence of events
that take place in the air and on the ground.

25

WHERE TO FIND BUTTERFLIES

You have probably spotted at least a few different species of butterflies in your life. They can be seen wherever the right plants grow, and are found on all continents, except for Antarctica. The highest diversity of butterflies is found in the Neotropical region, with about 40 percent of all species concentrated there. To give you an idea of how megadiverse the Neotropical butterfly fauna is, more than 1,300 species of butterflies were recorded at a single site of less than 15.5 square miles (40 square kilometers) in a study done in southeastern Peru. That is far more species of butterflies than can be found in the whole of the USA, a country covering more than 3.8 million square miles (9.8 million square kilometers). However, high species diversity does not necessarily correlate with high abundance.

Adult butterflies can be found in a great variety of different habitats ranging from degraded urban areas to pristine rainforests. All they need are their host plants for the caterpillars and flowers with nectar for adults, but, of course, some types of habitats can support more butterflies than others. There is a general correlation between plant diversity and butterfly diversity; in other words, vegetation with more types of plants supports more species of butterflies. Forest edges, and clearings and openings within forests, are great places to look for butterflies. There are many

BELOW | A bird's-eye-view of the rainforest at Tambopata National Reserve (Madre de Dios, Peru). This site has one of the highest number of species recorded for butterflies at a given site.

ABOVE | A butterfly trap in action. Traps are typically cylindrical and closed at the top to keep the butterflies inside once they have entered.

LEFT | A mown grass path in a meadow along the forest edge in the countryside in the UK. Forest edge is generally a great place to look for butterflies.

other factors that come into play (e.g., temperate versus tropical climate), but if you are looking for a specific species of butterfly, it is important to have some knowledge of its biology: What does it eat? What time of the year does it fly, and in what conditions or during what part of the day? For instance, not all butterflies feed on flower nectar; some feed on rotten fruit, tree sap, or animal droppings. The Poplar Admiral *Limenitis populi* (Nymphalidae) is one such butterfly, known to feed exclusively on these decaying substances. This gorgeous nymphalid has received attention from butterfly enthusiasts who use various ways to bait it, which involves being in the field at the right time in summer. There is even a hotel in Japan that provides its guests with baits to use when looking for Poplar Admirals in the area.

Butterfly photography has become more common in recent years and photographs contribute greatly to our understanding of the creatures. Nevertheless, our knowledge of butterflies relies heavily on a long tradition of collecting. Using a hand net is still the most popular way of collecting butterflies, but using baited traps is crucial to understand specialized groups or guilds, especially in the tropics, where many species live high up in the canopy of rainforests. Baits using rotten fish, seafood, or carrion are best for capturing a wide diversity of species and guilds, and certain species in various families are only attracted to baits of animal origin. Fruit attracts both sexes of a limited number of butterfly groups.

BIOGEOGRAPHY

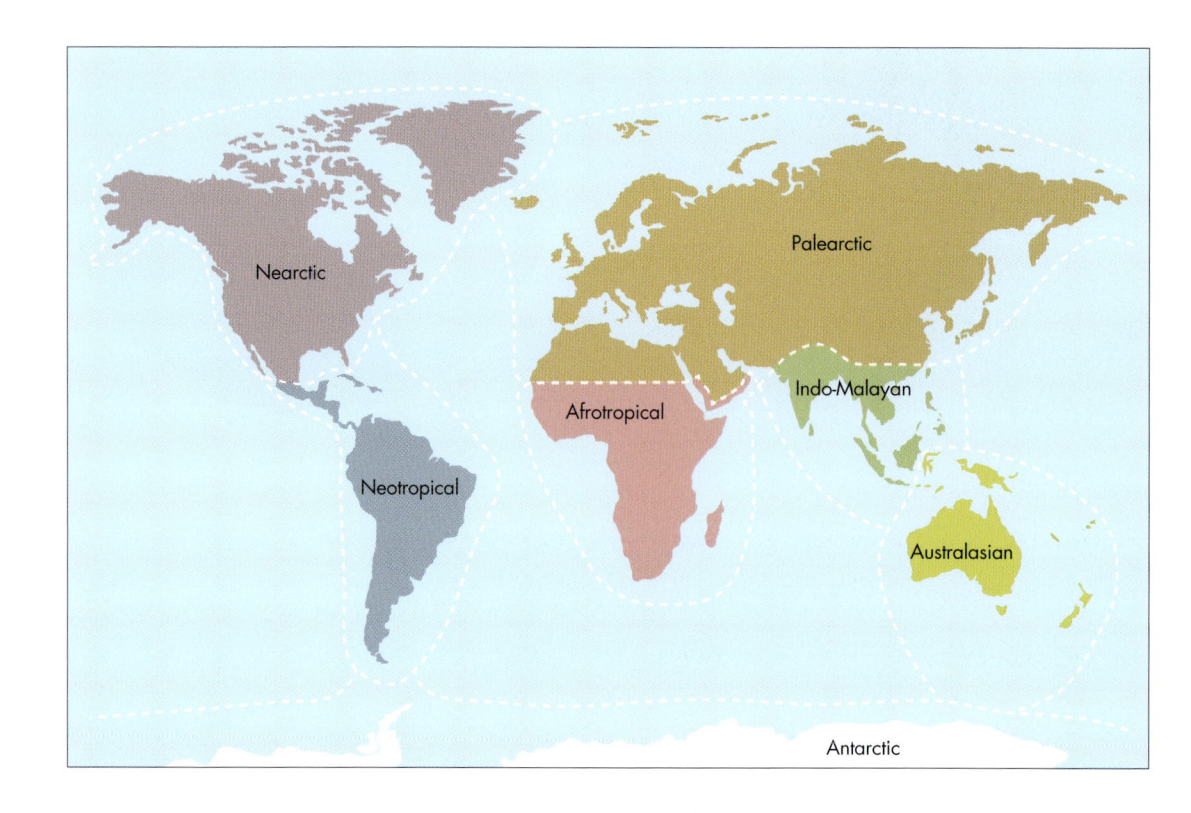

When considering the butterflies occurring in each of the six zoogeographical regions on Earth (Palearctic, Nearctic, African, Indo-Malayan, Neotropical, and Australian), it is clear that each region has its own characteristic butterflies. For example, birdwings in the genus *Ornithoptera* (Papilionidae) are only found in the Australian region, whereas members of a different genus of swallowtails, *Parides*, are restricted to the Neotropical region. On the other hand, the Painted Lady *Vanessa cardui* (Nymphalidae) has a cosmopolitan distribution and can be found in all of these zoogeographical regions. The theory of continental drift and plate tectonics contributes greatly toward explaining such distribution patterns. Additionally, increased use of molecular techniques and computational tools help us better

ABOVE | Map showing the zoogeographical regions of Earth.

to understand the biogeography of butterflies and provide insights into why many groups are found in certain parts of the world and not others. On the following pages the butterfly fauna of each of the zoogeographical regions are briefly summarized. There are no butterflies in Antarctica.

LEFT | A mating couple of Diana Fritillary *Speyeria diana*. This gorgeous fritillary exhibits a great degree of sexual dimorphism and is one of the more charismatic butterflies in North America.

BELOW | Members of the genus *Lepidochrysops* are only known from Africa.

HOLARCTIC REGION

The Holarctic region encompasses the Palearctic (Eurasia and North Africa) and the Nearctic (North America and Greenland). The Palearctic and Nearctic regions have many butterfly groups in common, and here the two are classified together under the Holarctic due to their faunistic resemblance to one another. The apollos (genus *Parnassius*) and relatives, the coppers (lycaenid subfamily Lycaeninae), and the fritillaries (nymphalid tribe Argynnini) are some of the butterfly groups that are diverse in this region and characterize the butterfly fauna. The majority of the Holarctic region falls within the temperate zone, and thus butterfly lifecycles are closely linked to climatic fluctuations associated with four distinct seasons (spring, summer, fall, and winter).

AFROTROPICAL REGION

This region refers to central Africa to South Africa, including Madagascar. Some diverse butterfly groups in Africa include the nymphalid genus *Charaxes*, the nymphalid tribe Acraeini, and the lycaenid tribe Leptinini. Over 600 species of Leptinini are endemic to continental Africa, and 85 percent of *Charaxes* diversity is concentrated in the continent. A notable feature of butterflies in Africa is that many distantly related species possess black spots around the base of their wings, which can be interpreted as mimicry. The highest diversity of butterflies on the continent is concentrated in West Africa, where one-third of African butterflies can be found. The butterfly fauna of Madagascar is distinctive with many endemic species such as the Madagascan Emperor Swallowtail *Papilio morondavana*. About 4,000 butterfly species are known from continental Africa and Madagascar.

INDO-MALAYAN REGION

This region encompasses the area south of the Himalayas and Southeast Asia. It is home to many distinctive and diverse tropical butterfly groups, such as the birdwing genus *Troides* (Papilionidae) and the nymphalid tribes Amathusiini and Adoliadini. The Indo-Malayan region is unique in the sense that it includes thousands of islands in Southeast Asia, generating numerous variations, forms, subspecies, and species restricted to one or few of these islands. In the mid-1800s, Alfred Russel Wallace studied swallowtail butterfly faunas on the islands in this region to infer area relationships. He referenced Bates's quote "[butterfly wings] *serve as a tablet on which Nature writes the story of the modifications of species*" in an article published in 1865, which helped to bring butterflies to the forefront of biogeographical research. Most of the Indo-Malayan region falls within the tropical zone; about 3,000 species of butterflies are known.

AUSTRALASIAN REGION

The Australasian region includes Australia, Tasmania, and New Zealand, as well as some islands to the north, such as New Guinea. The region is separated from the Indo-Malayan region by Wallacea (the area between the so-called Wallace line and Weber's line). The birdwings *Ornithoptera* are among the most spectacular butterflies in this region. The Queen Alexandra's Birdwing *O. alexandrae*, found in eastern Papua New Guinea, is the world's largest butterfly by wingspan—the female's exceeds 11 in (28 cm). In the nineteenth and early twentieth centuries, some foreign naturalists shot these butterflies with dust guns, leaving bullet holes in their wings. Tropical rainforests in New Guinea and adjacent islands are home to many other striking and diverse groups, such as the genus *Delias* (Pieridae). Peculiarly, skippers are absent from New Zealand. While New Zealand only has around 25 species of butterflies, the Australasian region as a whole is home to about 1,000 known

species. The true number is probably higher since the region includes many unexplored islands and remote areas. In fact, a spectacular new species, the Natewa Swallowtail *Papilio natewa* (Papilionidae) was discovered on a Fijian island in 2017.

NEOTROPICAL REGION

The Neotropical region extends from areas of southern USA and Mexico through Central and South America, and includes the Caribbean. The region is known to host many endemic and showy butterfly groups including the blue morphos and *Heliconius* butterflies. The moth-like butterflies (Hedylidae) are only found in the Neotropical region and more than 90 percent of species in Riodinidae are concentrated in the Neotropics. The unique butterfly fauna of the Neotropics is influenced by the Andes Mountains, which serve as a barrier separating the distinctive lowland faunas on each side, but the Andes themselves also contain elevational gradients and harbor distinctive and rich butterfly faunas, especially in the cloud forests. Further, Amazonia is the epicenter of Earth's biodiversity, and the world's largest tropical rainforest is home to thousands of butterfly species. New species are described from these areas every year.

The distinctive butterfly fauna of the Atlantic coastal forest, which stretches mainly along the southeastern coast of Brazil, adds another layer of diversity and uniqueness to the rich Neotropical butterfly assemblage. Due to these different biogeographical regions contributing to the overall Neotropical butterfly fauna, the region has the highest number of butterfly species of all the six zoogeographical regions. About 8,000 species of butterflies are known from this region, accounting for 40 percent of the global butterfly diversity.

SURVIVAL STRATEGIES

Living in nature is tough. Unfavorable environmental conditions arise, and predators and parasitoids are looking for prey. Adult butterflies lack obvious weapons to fight and are seemingly defenseless, yet they fly during the day when they are highly visible to potential predators. They tend to be more cryptic during their immature stages, when they are rather immobile compared with adults. However, butterflies diversified in many parts of the world and became abundant, perhaps owing to their various survival strategies, both as an immature and as an adult, as briefly introduced here. It is important to keep in mind that insects are experts at developing survival strategies!

MIMICRY

One of the best-known survival strategies of adult butterflies is mimicry. There are two types of mimicry: Batesian mimicry and Müllerian

mimicry. Batesian mimicry occurs where palatable species resemble chemically defended, unpalatable, distantly related species to avoid predators such as birds. In Batesian mimicry, the model species does not receive any benefits. Classic examples of Batesian mimicry are found among Neotropical butterflies, such as mimics in the genus *Dismorphia* (Pieridae) and their unpalatable model *Mechanitis* (Nymphalidae), and *Patia* (Pieridae) and their unpalatable model *Methona* (Nymphalidae).

These examples are discussed in Henry Walter Bates's article, *Contribution to an insect fauna of the Amazon Valley*, where he presented his theory of mimicry in 1862. We also see many cases of Batesian mimicry in other parts of the world: most species of palmflies of the genus *Elymnias* (Satyrinae) resemble various distantly related unpalatable model butterflies—mainly in the nymphalid subfamily Danainae, such as those in the genus *Euploea*—throughout Southeast Asia. The moth *Cyclosia midamia* (Zygaenidae) is reported to be involved in Batesian mimicry not just as an adult but also as a caterpillar. The caterpillar of *C. midamia* resembles an unpalatable swallowtail caterpillar, and the adult appears to be a mimic of the Striped Blue Crow *Euploea mulciber*.

Müllerian mimicry is an adaptation in which unrelated toxic species mimic each other to share the cost incurred during the course of predator learning. Members of the genus *Heliconius* might present the best example of Müllerian mimicry, demonstrated by unrelated *Heliconius*

LEFT | Plate showing some examples of "mimetic analogies" in the Amazon rainforest. Plate LV from Henry Walter Bates' article published in 1862.

species occurring in sympatry having similar wing patterns. All *Heliconius* species feed exclusively on *Passiflora* plants (Passifloraceae) and incorporate the toxins produced by plants to chemically defend themselves. These butterflies resemble each other and display warning colors to advertise their distastefulness to predators when on the wing.

CAMOUFLAGE

The Oakleaf *Kallima inachus* (Nymphalidae), found in tropical Asia, is famous for the resemblance of the underside of its wings to a dried fall leaf when the wing is closed. But when the wings are open it displays very contrasting iridescent blue and orange. Similarly, many butterflies classified in the nymphalid subfamily Satyrinae possess rather dull brown wings; they blend into surroundings such as tree bark or dead leaves when at rest. Pupae of some butterflies may change color depending on their location and this adaptation can be considered as camouflage. Swallowtail butterflies have two pupal colors, green and brown; studies suggest that green pupae on a green background are less likely to be found by predators. This "blending in" strategy can be found in caterpillars too as their textures and coloration typically match their host plants. Early instars of many swallowtail caterpillars resemble bird droppings, which seems to be a way of fooling birds into thinking that caterpillars are their own excrement (it is possible that they can fool parasitoid wasps too!). Additionally, swallowtail

a

b

c

d

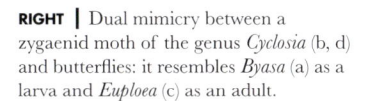
RIGHT | Dual mimicry between a zygaenid moth of the genus *Cyclosia* (b, d) and butterflies: it resembles *Byasa* (a) as a larva and *Euploea* (c) as an adult.

caterpillars possess a Y-shaped organ called the *osmeterium*, which pops out from behind the head and emits a volatile substance with an unpleasant odor when the caterpillar is disturbed. Skippers and some lycaenid caterpillars build shelters by rolling and stitching leaves in an attempt to hide from predators. The Orange Hairstreak *Japonica lutea* (Lycaenidae) rubs substances from the host plant (oak trees) and its own scales onto eggs after oviposition to make the eggs less visible. Similarly, females of the Atala Hairstreak *Eumaeus atala* (Lycaenidae) are known to cover their eggs with brightly colored scales from their abdomen; however, this creates an aposematic signal rather than camouflage.

CHEMICAL DEFENSE/MIMICRY

As briefly summarized in the life-stages section, caterpillars of some butterflies, mainly Lycaenidae and Riodinidae species, have symbiotic relationships with ants. This interaction and reliance on ants can vary depending on the life-stage of the butterfly and range from

to Lycaenidae and Riodinidae, observations of similar behavior by other butterflies exist, such as the Asian Skipper *Lotongus calathus* (Hesperiidae), whose caterpillar lives with ants inside its shelter.

OTHER STRATEGIES

Adults of many satyrines exhibit eyespots on their wings, hence their common name, ringlets. Larger eyespots on the middle of the wings, such as those of the owls (nymphalid genus *Caligo*), are thought to deter predators, whereas smaller eyespots along the margins of wings are thought to draw predators' attention away from vital parts of the body. A similar strategy of distracting predators is used by members of the hairstreak subfamily Theclinae. When adults of these lycaenids rest with their wings closed, the posterior end of their hindwings resembles a butterfly's head by having small eyespots (or markings) and a short tail, creating a "false head." The feigning of death, *thanatosis*, is commonly seen in species in the nymphalid subfamily Danainae, but it can also be seen in other nymphalids. When we try to handle these Danaid butterflies, they tuck their legs tight against the body and expose their forewings and bright hair pencils on their curved abdomen. Thanatosis is presumably a way of avoiding predator attack and preventing further damage to vital parts of the body.

facultative to obligate, as well as mutualistic to parasitic. Caterpillars are protected from predators and parasites by ants and, in return, reward the ants by providing attractive secretions from specialized glands in the cuticle. Among these species are some that are predatory, such as the Scarce Large Blue *Maculinea teleius* (Lycaenidae). Early instar caterpillars of the Scarce Large Blue are herbivorous, whereas mature caterpillars secrete chemical compounds to manipulate foraging worker ants into taking them into the ant nest. Once in the nest, the caterpillars continue to fool the ants with their chemical secretions, and then they start preying on juvenile ants. This manipulation via chemical compounds can be interpreted as chemical mimicry. Although reports on ant-caterpillar associations are mostly related

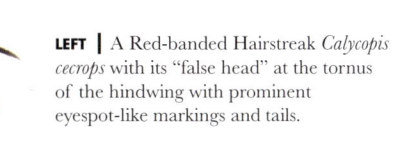

LEFT | A Red-banded Hairstreak *Calycopis cecrops* with its "false head" at the tornus of the hindwing with prominent eyespot-like markings and tails.

CONSERVATION

It hardly needs saying that we are losing biodiversity at an alarming rate, driven by deforestation, climate change, human population growth, and various human activities. A compilation of multiple recent surveys has shown that insect populations have declined by 45 percent over the past four decades, meaning that much of the so-called insect apocalypse has happened within the lifetimes of many of us. Unfortunately, we do not have any meaningful solutions for saving biodiversity immediately from this crisis, but we must keep trying. As we move further into the Anthropocene—the current period, during which human activities are having a significant impact on the planet—it is crucial to develop a better understanding of Earth's biodiversity as an essential prerequisite for learning how to conserve it. Although many butterfly species need conservation strategies developed for their own protection, their popularity with the public makes them effective "flagship" and "umbrella" species for engaging people and fostering support for conservation of the areas or habitats they represent—including other species found there. Butterflies are used as a proxy for all insects, and surveys of butterflies suggest an overall decline in insects. Therefore, butterfly conservation is important not just for its own sake but also to better understand and save the myriad species they share a home with, while there is still time. Sadly, we might have lost a number of butterfly taxa in our lifetime, which we don't even know about, and many more are at risk of extinction. It is important to emphasize that trying to reverse the decline of insects is not just a scientific matter, it requires a political and society-wide effort.

ABOVE | Bank notes and postage stamps featuring the Jamaican endemic Homerus Swallowtail *Pterourus* (*Papilio*) *homerus*. This is the largest butterfly in the Americas, and these bills and stamps will contribute to its status as a flagship species.

STRESSORS

Multiple stressors have been identified that have contributed toward the reduction in insects: global warming (including climate-induced changes in range, storm intensity, droughts, fire, deforestation), agricultural intensification (including nitrification and insecticide use), pollution, urbanization, and species introductions. However, it is often a combination of these factors that directly or indirectly influences butterfly populations, thus their interactions and the relative importance should be assessed on a case-by-case basis.

In temperate regions, there seems to be a consensus that butterflies that inhabit meadows and pastures are suffering from steep declines. These grassland species are vulnerable to changes in farming practices (e.g., intensification) and land abandonment. Grasslands in temperate regions used to be maintained by periodic cutting and burning, which prevented ecological succession

and maintained grasslands and prairie habitat. In the Midwestern USA, prescribed fires are important for controlling invasive plants in prairie fens, and the ecosystem is fire-dependent. However, it is often no longer desirable to continue to maintain grassland habitats using traditional practices. Also, there is a need for afforestation, which together would result in a great reduction in grasslands. Grassland butterflies have herbaceous host plants and visit flowers in meadows to feed; thus fragmentation of grasslands results in their decline. Additionally, deer grazing can severely reduce grassland butterflies as the deer graze plants right down to the ground. The nationwide explosion of the deer population in Japan drove several butterflies to the brink of extinction, including *Aporia hippia*

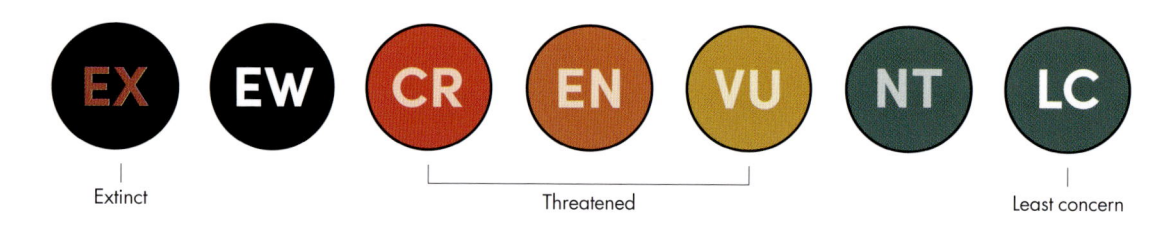

EX — Extinct

CR EN VU — Threatened

LC — Least concern

(Pieridae) and *Mellicta ambigua* (Nymphalidae). Although not properly documented, repeated over-collecting by multiple collectors at a single site in areas where range-restricted species occur can contribute to the decline of populations of certain species of butterfly.

IUCN RED LIST

The International Union for Conservation of Nature's (IUCN) Red List classifies imperiled organisms into nine categories according to their assessment of the risk of population collapse: Not Evaluated, Data Deficient (DD), Least Concern (LC), Near Threatened (NT), Vulnerable (VU), Endangered (EN), Critically Endangered (CR), Extinct in the Wild (EW), and Extinct (EX). Teams of experts working with the Sampled Red List Index have tried to assess about 1,500 species of butterflies (excluding subspecies and variations) and provided assessments as to their conservation status to inform decision makers. Of the randomly sampled species, fewer than 10 percent are currently categorized as Endangered or Critically Endangered.

More than half of the butterflies listed by the IUCN Sampled Red List are from the Afrotropical region, whereas a mere 7 percent are from the Neotropical realm. Considering that butterfly diversity is higher in the Neotropics, the number of butterflies appearing in the IUCN Red List is not proportional to the perceived species richness in these two regions. This discrepancy does

ABOVE | IUCN Red List categories: EX, Extinct; EW, Extinct in the Wild; CR, Critically Endangered; EN, Endangered; VU, Vulnerable; NT, Near Threatened; LC, Least Concern.

not mean that African butterflies are more endangered: it is a reflection of the poor state of our knowledge of butterflies in the Neotropics in general and the lack of a systematic program to assess threat for butterflies. For example, many butterflies in the Neotropics have not been recorded since they were described 100–150 years ago, and some species are represented only by a single specimen. In particular, the case of skippers (Hesperiidae) in the Neotropics shows how fragmentary our knowledge is: only two skippers from this region, currently, have been sampled out of the 128 Hesperiidae species listed.

In contrast, a comprehensive assessment of the conservation status of European butterflies was recently conducted (European Red List of Butterflies), with the survey suggesting that approximately 10 percent of European butterflies are under threat. Unlike for butterflies in temperate regions, due to a lack of information on life history and distribution data for many butterflies in the tropics, we are still far from the point where we can assess the conservation status of all butterflies appropriately. Collection of information available from resources such as research papers, species inventories, and curated occurrence data on online platforms, such as

GBIF and iNaturalist (global digital repositories for biodiversity data), combined with expert knowledge and scientific collections, will improve the coverage and effectiveness of the IUCN Red List assessments. Butterfly scientists, collectors, citizen scientists, and photographers should all be united in the effort to collate information that will help guide action toward avoiding and reversing butterfly declines. Proactive groups of researchers have launched national efforts to publish lists of threatened species ad hoc, such as the Red Book of Endangered Fauna of Brazil.

HOPE

Efforts toward conserving butterflies can be seen in the form of work by various nonprofit organizations and research projects conducted at academic institutions, all targeted at monitoring and saving this important group of insects, as well as the many habitats they rely on. Most of these conservation organizations are active in temperate regions, such as Butterfly Conservation in the UK and Europe, the North American Butterfly Association in the USA, and the Japan Butterfly Conservation Society in Japan. Due to the lack of institutional capacity to support biodiversity research in countries in the Global South, only a handful of groups are making contributions toward saving butterflies, but there is progress. While much of this section has focused on the decline of butterflies, climate change has influenced the abundance and range of some butterflies in temperate regions, especially those at the edge of their distributions. Even while some species are suffering losses due to change, such as to climate or land use, there is hope that some populations can adapt and survive with help from nature-loving people.

BELOW | Conservation effort in situ by the Japan Butterfly Conservation Society, a nonprofit organization aimed at saving endangered butterflies in Japan.

EXTINCTION

In *On the Origin of Species*, Charles Darwin recognized extinction as part of natural selection: "species and groups of species gradually disappear, one after another, first from one spot, then from another, and finally from the world." Indeed, we know that some butterflies have become extinct. Some of these are represented by a handful of fossils or specimens held in museum collections, while others are known only from illustrations.

Three of the many species of butterflies listed as Extinct on the IUCN Red List are *Libythea cinyras* (Nymphalidae), *Lepidochrysops hypopolia*, and *Deloneura immaculata* (both Lycaenidae). One of these, the Mauritius Snout *L. cinyras*, is known from a single specimen collected in Moka, Mauritius, in 1865. Roland Trimen published the description of this butterfly in 1866 and implied

that there was another specimen in the "South African Museum," but this specimen has never been located. The single known specimen is held at the Natural History Museum in London. Two African lycaenids were also last recorded in the late 1800s and are considered not to have been rediscovered since, although more research is needed to clarify the identities of these hairstreaks. It is important to emphasize that most tropical butterflies need their status assessed.

Fossil records and amber inclusions of butterflies (i.e., trapped in resin) are limited compared with other fossilized insects, and about

30 are known. Among these fossils, the nymphalid *Prodryas persephone* from the Eocene Florissant shale beds in Colorado is the best-preserved butterfly fossil. These fossils connect us to ancient life, as well as providing insights into the evolutionary history of butterflies.

In reality, there are many butterflies, especially in the tropics, that we do not know much about and are unidentified. These enigmatic butterflies are known only from illustrations, often accompanying the original description, while the specimens used to prepare the description by the author (type material) have been lost. An example is the Neotropical skipper *Papilio flavomarginatus* (Hesperiidae), which was named in the mid-1800s by Dutch entomologist Jan Sepp, who described many butterflies from Suriname based on observation and drawings made by a naturalist, H. J. Scheller. Scheller traveled to Suriname in the late 1700s and made life history notes and illustrations by directly observing immatures as well as adults. Sepp later acquired Scheller's field notes and drawings, made modifications, and added text to describe taxa he considered new to science, presumably without having physical specimens. At this point, we do not know of any skipper specimen that matches Scheller or Sepp's drawings. Another enigmatic specimen was described by Johan Christian Fabricius as *Hesperia busiris* in 1793 based

on a drawing prepared by William Jones. No specimen was known and for more than two centuries this species was classified as a skipper of an unknown origin. Recently, a noctuid moth, presumably from western Africa, described in 1854 was concluded as conspecific with *H. busiris* based on a single known specimen, which might have been used to name both "butterfly" and moth.

Pl. 142

HISTORY OF STUDYING BUTTERFLIES

Humans and butterflies presumably had close interactions when we developed agriculture about 12,000 years ago and transitioned into a more sedentary lifestyle from a nomadic way of life. As we had to grow and produce our own food, identifying and understanding pests and non-pests would have been a critical part of the farming lifestyle. We had to classify beneficial, neutral, and harmful insects to survive. This attempt to classify insects, which undoubtedly included butterflies, can be considered as an embryonic phase of taxonomy (the science of naming and classifying organisms); it is important to emphasize that we have been classifying objects since the beginning of humanity, and we will continue to classify them until we disappear from this planet. One of the first attempts to systematically classify organisms was made by Aristotle. He divided organisms into groups, one of which, termed *Entoma*, included insects, which he further classified based on the presence or absence of wings and mouthpart characters. A translated version of Aristotle's *Historia Animalium* suggests he examined butterflies (possibly the Cabbage White *Pieris rapae*) and observed the immature stages.

A notable figure in the study of butterflies is Maria Sibylla Merian (see p. 19). She was born in

BELOW AND LEFT | Portrait of Maria Sibylla Merian in her later years, and her painting of the Menelaus Blue Morpho *Morpho menelaus* and the Pomegranate Tree *Punica granatum* (Lythraceae). The caterpillar depicted is a Banded Sphinx Moth *Eumorpha fasciatus*, not a Menelaus Blue Morpho.

Germany in the mid-seventeenth century and grew up raising butterflies and moths. She dedicated herself to practicing her watercolor painting skills by meticulously documenting plants and their interactions with butterflies, caterpillars, and other insects. Her commitment and engagement in painting the metamorphosis of caterpillars led her to travel to the then Dutch colony of Suriname. Merian spent two years capturing the lifecycle of tropical butterflies with their host plants (as well as other creatures) through engravings and watercolors, which led to the production of her monumental work *Metamorphosis Insectorum Surinamensium*. Her attention to detail coupled with her artistic talents means that her work is prized for its aesthetic values as well as its scientific merit. Like the butterflies she painted, Merian and her work sit at the intersection of art and science. Because of its dual nature, her work was far-reaching and provided evidence of metamorphosis (although she was not the first to document it) at a time when many people believed in spontaneous generation.

Merian's work influenced Carl Linnaeus's *Systema Naturae*, in which binomial nomenclature was adopted (tenth edition, 1758), and where all known butterflies, from swallowtails to skippers, were classified under the generic name *Papilio*. Over 300 species of butterflies were described by Linnaeus, including several misidentified moths, such as *Urania*. Linnaeus' work significantly advanced our understanding of butterflies and other organisms by assigning a binomial to each of them as well as producing order by grouping them into nested hierarchies (class > order > genus > species). Subsequently, Linnaeus's student Johan Christian Fabricius established the genus *Hesperia* to accommodate those species known today as skippers. Fabricius was a prolific Danish entomologist known for naming about 10,000 new species of insects, including over 1,600 butterflies and moths. Fabricius further motivated Pierre André Latreille to publish his *Précis des caractères génériques des insectes, disposes dans un ordre naturel* (*Summary of generic characteristics of insects, arranged in a natural order*) in 1796 where he introduced the concept of family and tribe. Two butterfly families, Papilionidae and Hesperiidae, are attributed to Latreille and the root for other butterfly families accepted today can be traced back to around this era. Latreille's colleague at the museum in Paris, André Marie Constant Duméril

introduced the term *Rhopalocera* (as *Ropalocéres*) in 1823, which was used to scientifically refer to butterflies by many subsequent authors. *Rhopalocera* means "clubbed antennae," and, in contrast, moths were termed *Heterocera*, meaning "variable antennae," reinforcing the classic distinction between butterflies and moths based on antennae shape. Similarly, the terms *Diurini* and *Nocturini* were also used for butterflies and moths respectively.

The Victorian era saw serious interest from the public in natural history specimens, and exotic tropical bird and butterfly specimens, especially, received special attention and commanded good prices at auction houses. The so-called Stevens Auction Rooms at 38 King St., Covent Garden, in London was where various items such as plants, butterflies, and elephants were sold, including specimens collected by Alfred Russel Wallace and Henry Walter Bates. Their work was financially supported by Samuel Stevens who was partially in charge of the auction room. Here, *Agrias claudina godmani* (Nymphalidae) from Brazil reached a record price for an insect. Evidently, there was a market that could support Wallace and Bates's expeditions. Similarly, many other naturalists in Europe or America embarked on journeys to tropical rainforests and brought back specimens.

In the second half of the nineteenth and the early twentieth century, there was an explosion of interest in naming and describing new butterfly species, especially from European explorers and naturalists, mostly from the UK or Germany. Our understanding of butterfly species diversity improved significantly during this time. German

LEFT | Stevens Auction Rooms at 38 King Street, London. From Emily Allingham's *A Romance of the Rostrum*, published in 1924.

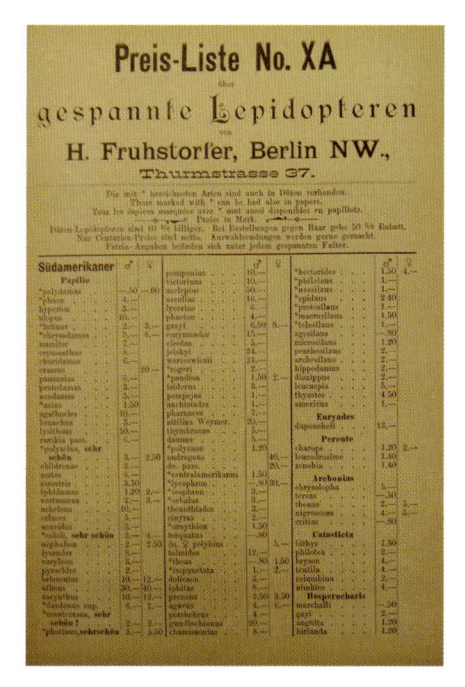

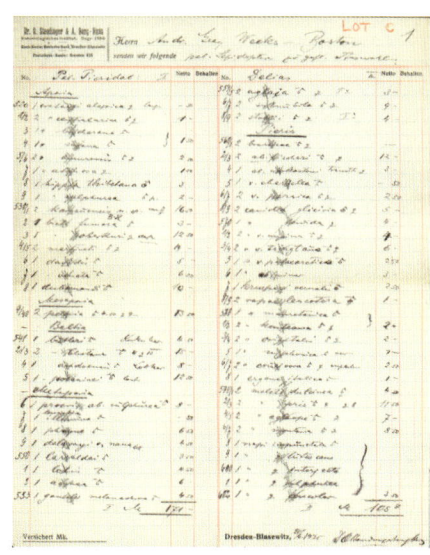

entomologist Hans Fruhstorfer was one of the most prolific authors, publishing descriptions of new butterfly taxa. Fruhstorfer traveled around the globe and supported himself by selling insect specimens and shells. The Dresden-based insect trading company Staudinger & Bang-Haas also issued catalogs and price lists for butterfly specimens and supported scientific endeavors. Otto Staudinger and Andreas and Otto Bang-Hass (father and son) were all accomplished Lepidoptera taxonomists and influential dealers of their time. Commercial venture and science seem to have been two sides of the same coin during this period. Walter Rothschild wrote in the preface of Emily Allingham's *A Romance of the Rostrum* (a story about the Stevens Auction Rooms), "I have always felt … the rooms at 38 King Street … were among the greatest aids and inducements to the study of Systematic Zoology." Like Stevens, Rothschild funded naturalists, including Albert Stewart Meek, who shot a Queen Alexandra's Birdwing *Ornithoptera*

alexandrae. The Victorian era was the time when science was not yet institutionalized, so many studies of butterflies were self-funded by naturalists who had a single-minded desire and enthusiasm to better understand butterflies, mainly by amassing specimens.

While the Victorian era cannot be experienced directly, the energy and enthusiasm of early naturalists can be felt through literature and archival records. The history of the study of butterflies has established the achievements of a diverse group of people, who contributed toward our understanding of these beautiful creatures, an understanding based on their innate motivation to explore the natural world.

STUDYING BUTTERFLIES TODAY

One major advancement in our understanding of butterflies has been achieving a certain degree of stability in classification owing to the DNA revolution. Seven families are recognized and their relationships seem well resolved at a higher classification level. The Darwinian era saw many scholars who influenced classification by promoting "tree-thinking" or "evolutionary thinking," such as Ernst Haeckel, building upon the groundwork of Latreille and Fabricius. The late twentieth century saw the foundation of Hennigian phylogenetics, which was followed by the introduction and subsequent rise of molecular techniques and computational power over the past few decades, elucidating the evolutionary history of butterflies. Consequently, we have now arrived at a more natural classification of butterflies, based on relatedness rather than resemblance. The more recent placement of the families Hedylidae and Hesperiidae in butterflies was supported by reconstruction of the phylogeny based on DNA sequence data. At the species level, genetic data have contributed particularly toward the discovery and description of cryptic butterfly species that were previously overlooked by experts.

In 2004, a group of researchers combined DNA data with caterpillar morphology and host plant records at a site in Costa Rica to show that skipper *Astraptes fulgerator* (Hesperiidae), previously thought to be a single common and variable species ranging from southern USA to northern Argentina is, in fact, a complex of at least ten species just in Costa Rica. DNA data also revealed that two satyrine species from South America, *Caeruleuptychia helios* and *Magneuptychia keltoumae* (both Nymphalidae), were in fact the male and female of the same species, morphologically distinctive due to sexual dimorphism. It has even been possible to extract and sequence DNA from museum specimens collected from the 1700s onward, so butterfly specimens housed at natural history museums and in private collections, all around the world, are now highly sought for DNA sequencing analysis. A whole new area of research has developed, termed *museomics*, meaning genomics dedicated to museum materials.

Trigonia

Celtis

Lonchocarpus

Inga and Cupania

Hampea

Capparis and Hampea

Byttneria

"Fabov"

Senna

Senna

LEFT | Diversity in caterpillar patterns that partly supported "ten species in one" in a study in Costa Rica of the skipper *Astraptes fulgerator* (Hesperiidae). Each name indicates the different plants that the caterpillars feed on.

Accumulation of butterfly specimens collected over centuries in natural history museums has allowed innumerable research projects to answer questions related to climate change, macroecology, evolution, and conservation. These existing specimens and their label data are being digitized and databased, and gradually being made accessible to the outside world, which will be used to answer some of society's pressing questions. At the Natural History Museum in London, the details of over 180,000 specimens of British butterflies collected over 200 years have been individually digitized. An international group of researchers used these data to assess the impact of climate change on the body size of British butterflies by investigating the relationship between body size and temperature. Their research suggests that adult butterfly body size increases with temperature during the late stages of larval development, providing insights into how butterflies can respond to climate change.

This type of research using large data sets coupled with international collaboration is typical of butterfly research conducted at academic institutions today. Further, with advancement of various technologies, we are able to look at certain aspects of butterflies in-depth. Scanning electron microscopy (SEM) is used for examining the ultrastructures of butterfly scales (see p. 15). A computed tomography (CT) scanning technique is an imaging procedure that can be used to capture internal structures. Researchers used CT scans to monitor the process of a butterfly pupa developing into an adult butterfly while still alive inside (Maria Sibylla Merian would have loved to have seen that!). With the development of artificial intelligence and machine learning, researchers have been able to systematically study evolutionary patterns in species diversity and compare males with females in some species, including birdwings.

ABOVE | These two specimens (upper- and undersides) were previously thought to represent two different species until DNA revealed that these are the male (top) and the female (bottom) of the same nymphalid butterfly, *Caeruleuptychia helios*.

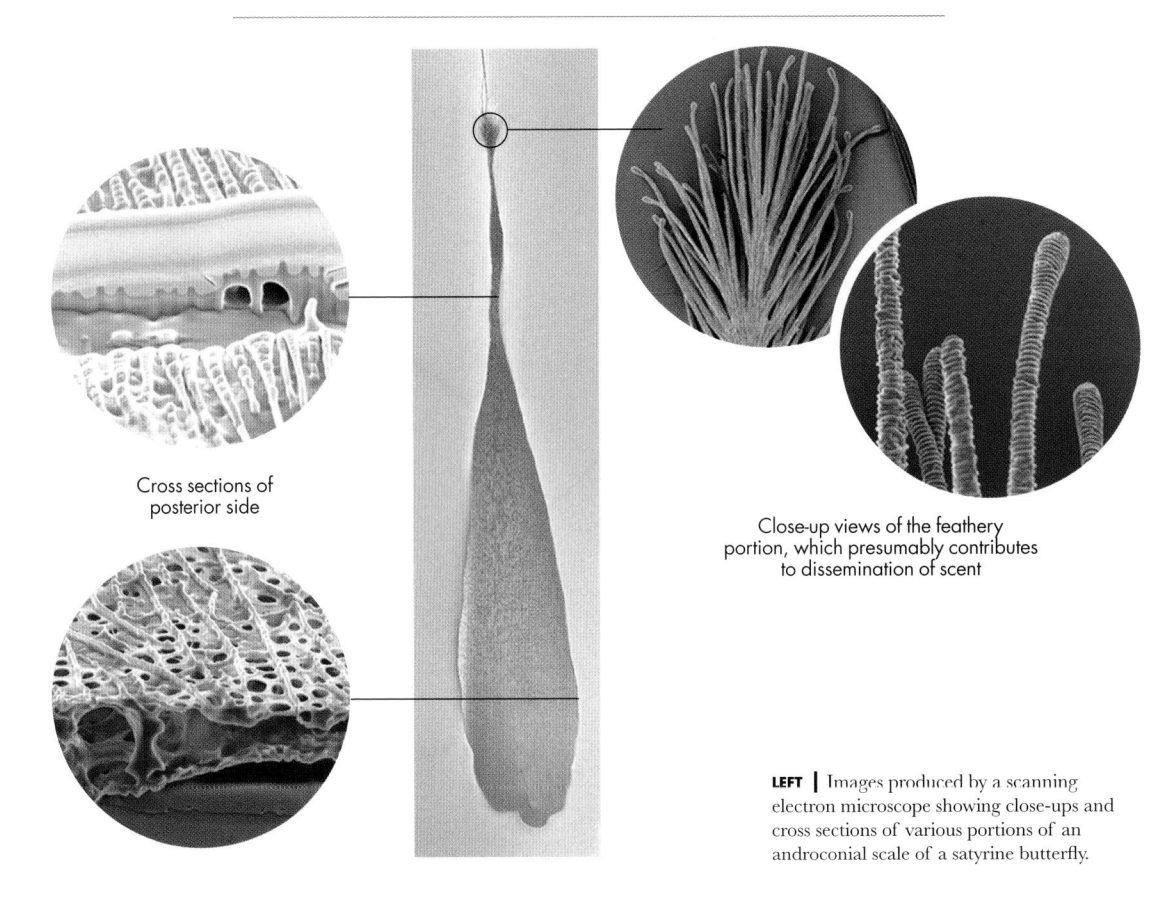

Cross sections of
posterior side

Close-up views of the feathery
portion, which presumably contributes
to dissemination of scent

LEFT | Images produced by a scanning electron microscope showing close-ups and cross sections of various portions of an androconial scale of a satyrine butterfly.

We live in a world where we rely heavily on the internet for obtaining information and for researching collaboratively. Additionally, many parts of the world can now be reached within a few hours or days. This can be advantageous for butterfly research, but if we care about the butterflies on the planet today we need regulations to prevent chaos. The Convention on International Trade in Endangered Species of Wild Fauna and Flora (CITES) is an intergovernmental agreement to ensure that international wildlife trading does not jeopardize existing populations of threatened species. Four butterfly species are considered endangered and receive the top level of protection under CITES Appendix 1 (swallowtails *Papilio chikae*, *Parides burchellanus*; Homerus Swallowtail *Pterourus* (*Papilio*) *homerus*; and the Queen Alexandra's Birdwing *Ornithoptera alexandrae*) along with rhinos, elephants, and gorillas. Specimens of these four species collected after July 1, 1975—dead or alive—are prohibited for international trade. Other species of butterflies are listed in CITES but under categories with fewer limitations. This does not mean that people can freely collect and study butterflies that are not listed under CITES. Many countries require collecting permits, even for studying common butterflies, in addition to export permits to take specimens out of their original country. Further, if the study involves DNA analysis, the Nagoya Protocol on Access and Benefit Sharing can come into play. This is an international agreement that came into force in 2014 and ensures that the benefits arising from the utilization of genetic resources are shared in a fair and equitable way (known as *access and benefit sharing*). The rationale behind this agreement is that genetic resources originating from a country must be shared equally with that country, particularly if they are used for commercial purposes.

The study of butterflies, especially natural history and taxonomic work, was undertaken by few museums and societies in Europe in the nineteenth century and relied considerably on the efforts of individual naturalists, such as Bates, Wallace, Rothschild, and Fountaine, to name only a few. However, nowadays it is becoming less simple for nonprofessional entomologists and naturalists to contribute to butterfly research due to the bar being raised and research depending on the use of molecular data and other advanced techniques that are not readily available to amateur entomologists. While they can continue to enhance current knowledge of butterflies, the limits to the types of work they can undertake and separation from professional entomology has its disadvantages. Modern science, founded on the study of the genome is based largely at institutions, but contributions from amateur naturalists and citizen scientists is crucial for a better understanding of the biology, ecology, and conservation of butterflies. Several online initiatives—including eButterfly and photography platforms such as iFoundButterflies, iNaturalist, Butterfly Catalogs, Butterflies and Moths of North America (BAMONA), and Flickr, and also social media platforms—show that there is a huge enthusiasm for butterflies. Since butterfly diversity is concentrated in tropical countries in the Global South, it is important to encourage and facilitate the inclusion of amateurs and researchers from various backgrounds in butterfly research.

CT SCAN OF PAINTED LADY CHRYSALIS

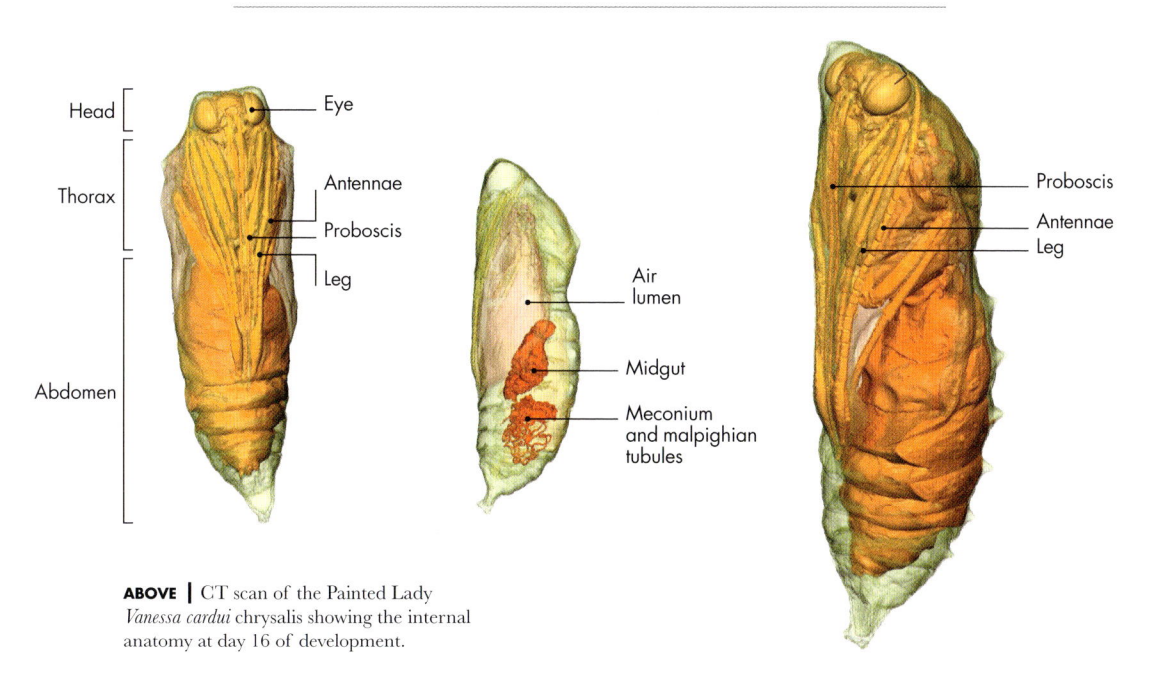

ABOVE | CT scan of the Painted Lady *Vanessa cardui* chrysalis showing the internal anatomy at day 16 of development.

BUTTERFLY NOMENCLATURE

SCIENTIFIC NAME VERSUS COMMON NAME

A butterfly represented by the scientific name *Papilio glaucus* is also known by the common English name Eastern Tiger Swallowtail, and both names refer to the same swallowtail, found in the eastern USA. Current taxonomy uses Linnaeus's method of naming species, called *binomial nomenclature*, whereby each species has a unique combination of two names, consisting of the genus name and the specific epithet. An advantage of this naming system is its inherent clarity—each taxon is known internationally by the same name—which is essential when fostering international communication about relationships among taxa. Hence, we can assume that *Papilio glaucus* is closer to other species in *Papilio* than to butterflies in other genera.

Common names vary according to language and even locally, which can lead to confusion. Common names can be informative and engaging if understood by the audience, but scientific communication is often between people who do not share the same language, so an accepted scientific name can be more useful. Most tropical butterflies do not have widely accepted common names. Conversely, some species have multiple common names, even in the same language. For instance, the butterfly known as the Mourning Cloak in North America and the Camberwell Beauty in the UK has one scientific name: *Nymphalis antiope*. Scientific names are meant to overcome this problem. Ideally, every species should be represented by a unique and universally accepted scientific name, although collisions do occur, where two different scientific names have been applied to the same butterfly or two different butterflies have been given the same name. The International Commission on Zoological Nomenclature (ICZN) published a set of rules for treatment and use of names to resolve such nomenclatural issues and many potential causes of confusion. These rules should be

TAXONOMIC NAMES

Category	Latin suffix (example)	Latinized suffix common name (example)	English common name
Order	-ptera	Lepidoptera	Butterflies and moths
Superfamily	-oidea	Papilionoidea (papilionoids)	Butterflies
Family	-idae	Papilionidae (papilionids)	Swallowtails, Apollo, and festoons
Subfamily	-inae	Papilioninae (papilionins)	
Tribe	-ini	Papilionini (papilionines)	Swallowtails

Genus and species names form a unique combination, for example, *Papilio alexanor*.
For subspecies a third name is added, for example, *Papilio alexanor destelensis*.

HOLOTYPE

Euptychia pillaca
Nakahara & Willmott

ECUADOR: *Zamora-Chinchipe*
km 10 Los Encuentros-El Panguí,
San Roque, ridge E
1050m, 3°42.11'S,78°35.36'W
4.viii.2009, K.R. Willmott & J. P.
W. Hall
FLMNH# 145783

Photographed by
K. R. Willmott 2011

DNA voucher
LEP-10711

LEFT | A type specimen usually bears a red label indicating that it serves as the standard for the species.

consulted before a name is selected for an undescribed new butterfly. Authors may name undescribed butterfly species after a characteristic feature of the insect, a place of origin, a person (or people), or in reference to an in-joke; ideally, it should sound euphonious. While not specified by the rule book, it is not recommended that you name a new species after yourself!

SUBSPECIES

Subspecies is the taxonomic rank below species. There are many subspecies recognized for butterflies. The subspecific name is added as a trinomen at the end of a binomen, and typically reflects a geographical variation in species occupying a wide range, although other naming criteria can be applied. Subspecies can be raised to species level and species can be downgraded to subspecies depending on the available scientific evidence. For example, in the Philippines, the Luzon Peacock Swallowtail *Papilio chikae chikae* is known from Luzon and another subspecies, *P. chikae hermeli*, from northern Mindoro. *Papilio chikae hermeli* was originally described as its own species, *P. hermeli*, but evidence suggests that it is conspecific with *P. chikae*, but represents a geographic variation. The Large Copper that became extinct in the UK in the 1850s is, in fact, the nominotypical subspecies *Lycaena dispar dispar* (Lycaenidae), while two other subspecies of *L. dispar* are extant in Europe and Asia.

TYPE SPECIMEN

Every new species described should have an objective standard. The specimen serving as this standard should represent the main characteristics that separate the newly named species from others. The objective standard, which shows these characters, is called the type specimen, and it is designated by the original author(s) or by subsequent designation. Modern descriptions include the designation of a holotype (one specimen), while in the past the concept of a species was typically based on a series of specimens (syntypes). The type specimen is a reference that can be checked if there is doubt about the identity of a species; thus it is important to mention in the paper describing it which institution holds the specimen so other researchers can locate and inspect it. A red or brightly colored label is frequently attached to type specimens to indicate their unique status. As science progresses, the interpretation and the concept of species can change, but the type specimen will remain as long as the specimen exists.

DISCOVERING NEW SPECIES

Discovering a new butterfly species is exciting. There are hundreds of butterflies still waiting to be discovered, described, and named by science. So where will these new butterfly species be found? Remote, unexplored areas mostly in the tropics are certainly good places to find new species. However, new species can be hiding in plain sight. Sometimes we just need to study our local butterfly fauna in-depth. For example, as recently as 2014, a new species, Intricate Satyr *Hermeuptychia intricata*, was found in the southern USA. Scientists often find new species when sorting out drawers in museum collections, where specimens collected at different times and from many different locations are stored.

So, what happens when someone thinks they might have found a new butterfly species?

The first step to check is the identity of the butterfly and that it has not been named previously. Typically, this is done by reviewing published literature and, if there is any doubt, checking type specimens and other specimens of related species to assess whether there are wing pattern or other variations. It is important to exclude the possibility of the butterfly representing merely a variation, a form, or an

BELOW | Intricate Satyr *Hermeuptychia intricata* resting on a leaf. The recent discovery of this species probably represents a proverbial "tip of the iceberg" of cryptic species hiding in plain sight.

aberration, of a known species. Discussion with butterfly taxonomists can also be beneficial. To characterize the species, the genitalia should be inspected for both sexes if available and/or the DNA sequenced to find which taxa it is closely related to and which genus it should be placed in.

Once the name has been chosen for the new butterfly, a paper fully describing it must be written and published in an appropriate journal to introduce it to the scientific community and to the world. The article should include figures illustrating notable features, especially the genitalia, as well as the body and wing patterns, habitats, host plants, and all the other information the authors have. The paper must be peer-reviewed by fellow specialists to be published in a scientific journal, fulfilling requirements of the ICZN. Once published, the name of the new butterfly species can be used officially to communicate about that butterfly.

BELOW | Vladimir Nabokov is known as a novelist and poet, but he also studied butterflies (especially Lycaenidae) at the Museum of Comparative Zoology (Cambridge, USA). As a taxonomist as well as a poet, he expressed his passion for naming new species in a poem "On Discovering a Butterfly".

BUTTERFLIES AND SOCIETY

The first butterflies emerged around 100 mya, while humans have been on the planet for just a few hundred thousand years. The earliest humans would have lived in a very different environment from ours today: closer to nature, sensitive to the changes in seasons, and closer to plants and animals, including butterflies.

Human attitudes to insects might not always be positive, but butterflies often stand apart. Most insects are seen as "creepy crawlies," with little favorable public attention. Insects are often thought of in terms of pest control to protect crops or reduce the transmission of diseases such as malaria. This was not always the case, though; insects in past cultures were revered (e.g., by the Egyptians and Mayans), but with increasingly intensive agriculture and an awareness of their pathogenic associations, they are often considered negatively (though their importance as pollinators is becoming more widely understood). Butterflies are an exception, although some species can be

pests of crops, being among the few insects that bring joy to people, due to their beautiful colors, delicate nature, and elegance.

Among insects, butterflies are one of the more conspicuous and charismatic groups, engaging the public's sympathy and interest. Butterflies attract attention as symbols of sunny days, freedom, and purity, and may also indicate a healthy environment. Some ancient cultures considered them to be the souls of the dead flying to heaven. Their day-flying habit and size make them relatively easy to observe, photograph, and paint, and they can also be bred relatively easily by amateur naturalists. Perhaps for all these reasons, the relationship between humans and butterflies is long and deep, and pervades science, art, and broader culture. Artistic interest emerged

BELOW | Early butterflies (and birds) depicted in an ancient Egyptian wall painting in the Tomb of Nakht, Thebes, Eighteenth Dynasty.

LEFT | Mural in Colombia of Literature Nobel prize winner Gabriel Garcia Márquez, author of *One Hundred Years of Solitude*. In the novel, the yellow butterflies swarm constantly around the main character, Mauricio Babilonia, as a symbol of magical realism.

BELOW LEFT | *Still Life with Flowers in a Decorative Vase*, c. 1670–1675, painted by Maria van Oosterwyck.

independently in various ancient human cultures. Some of the oldest depictions of butterflies can be seen in geometric designs in Scottish Neolithic stone carvings (c. 3000–5000 BCE). Truly identifiable butterflies appear in Egyptian times, with examples in tomb drawings, amulets, and jewelry from the Old Kingdom period (c. 2686–2181 BCE) onward. The tomb of the high-ranking official Nebamun (1350 BCE), in particular, contains vivid illustrations of butterflies in a hunting scene, providing evidence of the presence of various species at a particular location at this time. They feature in gold ornaments from pre-Columbian cultures of South America too.

In the Renaissance, classical, and modern art periods, butterflies were included in paintings as symbols of beauty, nature, freedom, resurrection, transformation, and life. They are found in many

artworks from these eras, especially where scenes are set in nature. The Dutch Golden Age painter Maria van Oosterwyck went further, using butterflies as the focus of her paintings.

Butterfly motifs are now widespread in art and jewelry and on clothing. They are frequently depicted on stamps, coins, and notes, with endemic or local species often depicted.

Butterflies have also been used as metaphors in literature. In Shakespeare's *King Lear*, the protagonist reminisces about butterflies. J. R. R. Tolkien describes an interaction with colorful butterflies as they fly in the sun in a treetop in *The Hobbit*, contrasting them with the lost explorers' experiences in the dark forest below. *Alice's Adventures in Wonderland* features a talking caterpillar, who later metamorphoses. In a poignant moment in the anti-war classic *All Quiet on the Western Front*, the butterflies in no-man's land contrast with surrounding desolation. And in *One Hundred Years of Solitude*, by Gabriel García Márquez, a virtuous character is followed by yellow butterflies wherever he walks. Márquez, who went on to win the Nobel Prize for Literature

following this work, hails from Colombia, the world's most biodiverse country for butterflies. In movie adaptations of all these books, and in other feature films such as *Encanto*, *The Butterfly Effect*, *Papillon*, *Bambi*, *Mary Poppins*, and more, butterflies often depict joy, color, and the innocence of nature.

The contemporary artist Damien Hirst has used butterflies in new ways. *In and Out of Love*, exhibited in 1991, featured live butterflies, which emerge from chrysalises, fly, feed, and die within the exhibit. In *I am Become Death, Shatterer of Worlds*, Hirst used 3,000 sets of butterfly wings in a kaleidoscopic image that has the appearance of stained glass.

Although most interactions between butterflies and people are positive, there are exceptions. Some caterpillars are pests. For instance, *Pieris rapae* (known as the Cabbage White in North America and Small White in Europe) can decimate crops of cabbage, broccoli, and brussels sprouts. The current rate of destruction of primary forests, where butterflies are most diverse, and increasing intensification of agriculture are now resulting in a so-called "Insect Armageddon", with butterflies affected just as badly as other groups. Fortunately, though, butterflies have increasingly become a focus for insect conservation.

The rise of ecotourism has increasingly involved butterflies. With the decline of nature in many areas of human habitation, butterflies can now be brought to the people in butterfly houses, spaces controlled for temperature, light, and humidity, allowing colorful and larger tropical species to be bred in the middle of cities and then to fly around fascinated visitors. Butterfly houses are largely commercial ventures, but as they become more popular in tropical countries, efforts are made to host local species, and so these initiatives are contributing to our knowledge of butterfly life histories, as well as raising awareness of conservation issues and bringing ecotourism income to local communities. Visitors to the Amazon or other tropical regions marvel at the diversity of butterflies, which are often a focus for photographers. A small but increasing number of butterfly tour operators take clients to see or photograph butterflies or to witness events such as the Monarch migration in Mexico. These initiatives bring increasing work opportunities for local butterfly experts and citizen scientists, and encourage local people to protect their natural resources.

Of all the insects, butterflies have, therefore, gained particularly prominent societal recognition.

BELOW | A live butterflies house in the Botanical Gardens in Quindio, Colombia, where the public can connect and engage with nature and its conservation.

BUTTERFLY CLASSIFICATION

The true butterflies, or Papilionoidea, are a superfamily (another formal rank in the taxonomic hierarchy) within the order Lepidoptera, which also includes moths. Traits that all the Lepidoptera share include scaled wings as adults, and the lifecycle of egg, caterpillar (larva), chrysalis (pupa), and adult (imago). Butterflies are a relatively recent grouping, and the monophyly of the superfamily, meaning that they descend from a common ancestor, has been widely accepted. The diurnal butterflies are currently arranged into seven families as listed on pages 10–11. There are 41 subfamilies in total, reflecting the vast array of guilds but also close relationships. Some of these are long-recognized and have always been considered as "butterflies."

Papilionidae (swallowtails) are a striking group of generally large butterflies, often with tail streamers. The Lycaenidae (blues, coppers, and hairstreaks) have an incredible diversity, not only in taxa but also in their lifecycles, which for many taxa are linked to ants (see pp. 159 and 161). Two groups that have been less studied are the Pieridae (whites and yellows) and the metalmarks in the Riodinidae family, a diverse group of small butterflies distributed around the world, though most are tropical.

The most species-diverse group is the Nymphalidae (brush-foots). These include various well-defined groups that are sometimes recognized at family and subfamily level, and sometimes recognized at the tribe level, including the Satyrini (ringlets), Morphini (morphos), and the Brassolini (owls); they also include some of the most familiar butterflies, such as the Peacock, Monarch, and fritillaries. Until recently two groups were excluded.

ABOVE | The Papilionidae family includes some of the most spectacular and large butterflies, for example, the Queen Alexandra's Birdwing *Ornithoptera alexandrae*.

Hesperiidae, the skippers, were often considered to be "sisters" to the butterflies and placed alongside them, or alternatively as a subgroup of moths. Molecular studies have shown the skippers to be part of the butterfly phylogeny. Finally, members of the Hedylidae are small and dull colored and were considered moths until

TOP | Museums around the world facilitate access to millions of specimens collected over hundreds of years, helping the study of butterfly classification.

ABOVE | Despite the fact that butterfly classification started more than 250 years ago, some remote areas still have species to be discovered. Recently, this new species, the Yariguies Ringlet *Idioneurula donegani*, was found on an isolated mountain in Colombia.

very recently, but are now classified as a small group of butterflies.

The total number of species of butterflies is uncertain as, with new research, technology, and communication tools available, new species are regularly described from the most diverse areas of the planet, especially the tropics. Disagreements among taxonomists and the rapid changes in nomenclature make it almost impossible to provide an accurate figure. There are approximately 20,000 species recognized and certainly some hundreds yet to be named. With the current unprecedented rate of extinctions of species on the planet during the Anthropocene, organizing and naming species has become urgent, in order to understand the relationships among organisms and to become aware of the existence of species that are not yet known but are at risk.

ABOVE | The Blue Morpho *Morpho helenor peleides*, a beautiful butterfly found from Mexico south to Argentina.

HOW TO USE THIS BOOK AND INTRODUCTION TO TAXON PROFILES

This book provides an overview of the currently recognized families, subfamilies, and tribes of the true butterflies of the superfamily Papilionoidea. The order in which the families are presented is intended to reflect their time on the planet, starting with the most ancient and oldest extant lineages. At subfamily and tribe levels, the arrangements generally follow the same structure, with those placed together being those with the closest relationships. However, some subfamilies and tribes are grouped under one profile, either because they are newly classified and we have less information about them, or for reasons of space. The number of pages devoted to each profile is determined partly by how much or how little is known about the group being described, and partly by the number of species in the group.

The classification and taxonomic arrangement presented here reflects an attempt at an up-to-date consensus, and reflects the most recently published taxonomic works and phylogenies. Most, if not all, of the arrangements are supported by molecular analysis of genetic material (DNA). We have then incorporated our experience gained over decades of studying butterflies in the field and in large museum collections around the world. Advances in detailed morphological and molecular analysis will doubtless continue, bringing further dynamic changes in classifications across the higher taxa. Current technology has accelerated the rate of discovery of new species, and the understanding of the relationships among them; this will inevitably cause changes to the number of species and genera recognized here.

Each profile begins with the current taxonomic status of the group being described; representative English names of the group are given. The main text then gives a summary of the group and how it is made up, an impression of the numbers within the group, and its main characteristics. Interesting features and key species are picked out, and some information about caterpillars and pupae may be given.

Each information panel begins with a range map, which is generic and provisional; the maps are based largely on distributional records found in the literature and some of the websites listed on page 234, as well as from museum collections. The distribution of each group is also summarized in the information panel.

The genera listed in the information panel are comprehensive for most groups; however, for those groups with the highest diversity, only a few representative genera are listed.

To give an idea of the size of the species in each group, the average forewing lengths (see below) are arranged into three categories: large (1 ½–4 in/40–100 mm), medium (¾–1 ½ in/20–39 mm), and small (<¾ in/<20 mm).

The "Host Plant Families" section includes some of the known plant families on which caterpillars of each group feed.

Photos have been chosen to show features typical of each group and give a flavor of the extraordinary diversity of wing patterns that butterflies display.

Forewing length (FW)

ABOVE | Forewing length as given in this book is measured as the longest straight-line distance from the wing base to the wingtip.

PAPILIONIDAE
SWALLOWTAILS, APOLLOS, AND FESTOONS

The family Papilionidae includes butterflies that are often large and colorful, and, despite having around 614 species, more than 2,086 subspecies, and 31 genera, it is the least species-diverse group of true butterflies after the moth-like family Hedylidae. Extant species are arranged in three subfamilies: Baroniinae, Parnassiinae, and Papilioninae. Members of a further subfamily, Praepapilioninae, are known only from fossils. Size, color, and habitats vary considerably in this cosmopolitan family, which can be found virtually everywhere from forest to deserts and from mountains to gardens. Extensively studied by scientists, Papilionidae is the best taxonomically documented group of butterflies, and probably of Lepidoptera; the whole family has benefited from being the only one fully assessed for threat in the IUCN Red List. The charismatic appearance of species in this group has encouraged the attention of amateurs and collectors over time, resulting in very large numbers of individuals stored in private and scientific collections around the world. This family is the most ancient group of butterflies, some of which were present on the planet millions of years before humans, even including some species that became extinct long ago. Recent molecular studies suggest that the subfamily Baroniinae is so ancient that it might one day be recognized in a separate family of its own.

ABOVE | A female
Great Mormon
Papilio memnon
(right) feeding on
nectar from flowers
and sharing with
another species of
swallowtail
butterfly.

Adults can be recognized from a combination of characters that can be identified by the amateur: the first (front) pair of legs are large and fully operational in both males and females and have well-developed claws; the antennae are relatively close at the base; and the hindwings often have projections like "tails" with their inner margins containing modified scales in males. However, these wing characters are not universal: they are absent in some species and can also be present in some species in other families. To the specialist eye, the arrangement of the veins, including a transverse nervure in the forewings, is key to distinguishing them from other butterfly groups.

Eggs are spherical and most do not have many patterns on the surface. The first instar caterpillars have long spines and later develop a retractile and bifurcated organ with a gland (osmeterium) that emits foul odors in response to disturbance. The pupa or chrysalis is angular and held upright by the cremaster, supported by tail hooks.

Parasites, natural enemies (predators), microclimate, and availability of the caterpillars' host plants are the primary factors that regulate populations of swallowtails. Practices affecting their host plants, such as destruction, conversion, and reduction of natural habitats because of deforestation, logging, crops, cattle grazing, and urbanization, are the biggest threats to all butterflies. However, iconic species have been instrumental in the development of a conservation culture, from butterfly houses engaging the public and educating them about tropical species and threatened tropical forests, to teaching children to appreciate insects. Some Papilionidae have become flagship or keystone species, envoys of campaigns for the protection of habitats and other less charismatic organisms, raising local, national, regional, and even international awareness. Species in the family Papilionidae have been fundamental study models for understanding evolution, genetics, mimicry, biochemical analyses, and, more recently, climate change.

ABOVE |
A spectacular Krishna Peacock *Papilio krishna*, a species found in forests in South and Southeast Asia.

THE LIVING FOSSIL BUTTERFLY

ABOVE | The Mexican endemic *Baronia brevicornis* is an enigmatic butterfly that has long lived on the planet but is currently under threat because of the loss of its habitats.

The most primitive species—and the oldest extant lineage of the true butterflies—evolved 80–90 mya and has a subfamily of its own. It is represented by a single genus and species, *Baronia brevicornis*, endemic (not found elsewhere in the planet) to dry forests in Mexico. Two subspecies are currently recognized: the nominate *brevicornis*, which was described in 1893, and the subspecies

GENERA
Baronia

DISTRIBUTION
Neotropical; endemic to southern Mexico

HABITATS
Local in deciduous scrub forests dominated by acacia-like trees at 1,600–4,500 ft (500–1,400 m) elevations in southern Mexico; seasonally dry tropical biome

SIZE
Small to medium: up to 1 ⅓ in (35 mm)

HOST PLANTS
Thorny bush *Vachellia campechiana* (Fabaceae) (synonym *Acacia cochliacantha*)

CONSERVATION
Categorized as Endangered on the IUCN Red List because of the fragmentation of habitats where the butterfly and host plant are found. Mature acacia trees are associated with grazing systems, and some habitats where *Baronia brevicornis* is found

rufodiscalis, named over a century later. These "living fossils" have a very distinctive appearance among Papilionidae, with dark, tail-less wings and scales colored yellow, orange, and brown. However, there are several forms (polymorphism) in both males and females, with around 25 "forms" named. Females can be larger, paler, and rounder winged than males. Males display strong territorial behavior.

The ecology of this species was unknown for over a hundred years until Mexican researchers pioneered detailed studies of its lifecycle and monitoring of the species populations. Eggs are laid individually, well-separated on the underside of leaves of bushy host plants, hatching after five days. Chalcid wasps can parasitize them. Caterpillars

develop five different instars in one month and build tubular structures from the leaves of the host plant. Apparently, this behavior reduces predation, as does the strongly odored substance secreted from a specialized structure called the osmeterium. The pupa develops underground, emerging as an adult after almost 11 months. Peak rainfall periods and humidity are environmental factors known to determine the emergence of both adult butterfly and the leaves in the host plant. Although adults may be found in various patches of forest, reproduction occurs only in areas with good availability of the host plant, where it comprises at least 70 percent of vegetation cover.

have been degraded by 40 percent due to agriculture, including cattle grazing, and timber extraction. Despite its localized distribution and IUCN status, this species is not listed by CITES. The EDGE program for species that are "evolutionarily distinct and globally endangered" does not include butterflies; however, this iconic species would be a prime candidate for listing

TOP LEFT | A male of *Baronia brevicornis* resting on plants. Its antennae are relatively short compared with other butterflies in the family.

TOP RIGHT | Females of *Baronia brevicornis* are distinctive and display different forms and colors.

PARNASSIANS, APOLLOS, AND FESTOONS

BELOW | The striking Spanish Festoon *Zerynthia rumina* can be found in rocky habitats and warm, dry areas in the Iberian Peninsula, southeastern France, and northern Africa.

The Parnassiinae subfamily includes medium- to large-sized butterflies with striking patterns in their wings. They have a remarkable distribution, flying in some of the most remote and highest elevation spots of the planet, such as the Himalayas. There are eight genera, almost 80 species, and over 350 subspecies recognized,

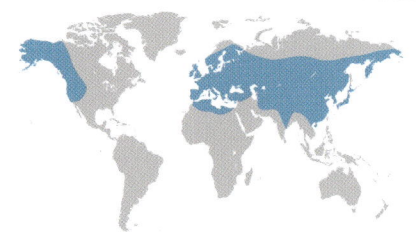

GENERA
Archon, Hypermnestra, Parnassius, Allancastria, Sericinus, Zerynthia, Luehdorfia, and *Bhutanitis*

DISTRIBUTION
Palearctic region in Europe and North Asia; Middle East, North Africa, Southeast Asia, and western North America

HABITATS
In Northern Hemisphere, most species in mountain habitats often at high elevations

and sometimes even up to 16,000 ft (5,000 m). *Hypermnestra* is found in arid deserts, *Luehdorfia* in humid forests, and *Zerynthia* in lowland meadows and mountains

SIZE
Large to very large: 1¾–4¾ in (45–120 mm)

HOST PLANT FAMILIES
Aristolochiaceae (*Archon, Allancastria, Sericinus, Zerynthia, Luehdorfia, Bhutanitis*),

arranged in two tribes, although some authorities recognize three tribes with the addition of the Luehdorfiini. Butterflies in Parnassiinae have a single anal vein in the hindwings, and the pretarsal claws of males are asymmetrical. Species in the genera *Parnassius* and *Zerynthia* show spectacular geographical variation that can be confusing; in the *Parnassius* genus alone, around 280 subspecies have been described and many more "forms." Members of these genera have been subdivided by some into

Zygophyllaceae (*Hypermnestra*), and Crassulaceae (*Parnassius*)

CONSERVATION
Bhutanitis ludlowi is currently categorized as Endangered on the IUCN Red List and listed in CITES Appendix II, along with other species in the genus: *B. lidderdalii*, *B. thaidina*, and *B. mansfieldi. Parnassius apollo* and many of its subspecies are also CITES listed

TOP | The Apollo butterfly *Parnassius apollo* is a good flyer over long distances. It shows variation in the eyespots and marks on its wings that can vary in size and number and also in color between individuals, but also between populations from different regions.

ABOVE | Some populations of the Apollo butterfly *Parnassius apollo* occupy small areas, including isolated mountains. The loss of habitat is undoubtedly the biggest threat for the long-term survival of the Apollo butterfly.

innumerable variations and forms, causing taxonomic inflation and confusion. Researchers using molecular characters have suggested that species in *Parnassius* occurring in remote areas of Central and Western Asia show morphological evolution, involving processes of loss and reappearance of characters such as red spots in the wings.

Because butterflies are ectotherms, regulating their body temperature using external sources, they are highly sensitive to their environment. Some species that fly at high elevations and species endemic to mountains have been excellent study organisms for climate

change research, especially the Apollo butterfly *Parnassius apollo*. The beauty and rarity of many Parnassiinae also make them sought after by scientists and collectors, though many species are now protected by international treaties and national laws.

Habitat degradation and loss are known to be the main causes for decline in mountain species. Populations of the endangered species *Bhutanitis ludlowi* were only recently rediscovered, 70 years after being initially found. It is now officially designated as Bhutan's national butterfly and protected there under strict laws.

Bhutanitis mansfieldi is only found in China. The rare and spectacular genus *Luehdorfia* includes various national endemic species such as *L. japonica* (Japan), and *L. taibai* and *L. chinensis* (both China). The Spanish Festoon *Zerynthia rumina*, popular because of the stunning display of dazzling colors in the upper and underside of its wings, is

nationally protected in Spain; it is also found in southeastern France and northern Africa.

Immature stages are commonly studied because of the display of several biological defense mechanisms, such as mimicry and camouflage. Caterpillars produce a repulsive taste for predators through chemicals obtained from host plants, which can be poisonous to vertebrates. They pupate inside cocoon-like webs, usually constructed among leaves and other debris. Recent studies of *Sericinus montela* caterpillars indicate that this species is multivoltine (having several broods a season), the only such case in the Parnassiinae subfamily.

SWALLOWTAILS

OPPOSITE | Species in the genus *Teinopalpus* possess a unique arrangement of photonic crystals in the wing scales, resulting in the striking matte green color.

BELOW | Known as Kaisar-i-Hind (Emperor of India), the stunning *Teinopalpus imperialis* has a very strong and rapid flight, and is found in pristine forests in high mountains in very few countries in Asia.

This is the largest subfamily in Papilionidae, and it includes the biggest and, probably, most spectacular diurnal butterflies. Without a doubt, it is the butterfly group that has received the most attention, with unavoidable and near-constant changes in its taxonomy. Its common name, swallowtails, is given because of the resemblance of the extensions in the hindwings to a swallow's tail, although not all species display this feature. Papilioninae is the only subfamily with representatives flying on all continents except Antarctica; its greatest species diversity is on the African and Asian continents. There are about 19 recognized genera, 532 species, and an impressive 1,700+ subspecies arranged in four tribes: Teinopalpini, Leptocircini, Troidini, and Papilionini. Life histories, behavior, and ecology of Papilioninae species are relatively well-studied. The range of host plants for this subfamily, although as vast as its number of species, is well-known for many groups.

GENERA
Teinopalpus

DISTRIBUTION
Southeast Asia: tropical and subtropical forests in China, India, Nepal, Laos, and Vietnam

HABITATS
Evergreen forests at mid-elevations up to 5,500 ft (1,700 m).

SIZE
Large: 2–2⅓ in (50–60 mm)

HOST PLANT FAMILIES
Magnoliaceae and Thymelaeaceae

CONSERVATION
Both species are included in CITES Appendix II although they are categorized as Near Threatened in the IUCN Red List. *Teinopalpus aureus* is listed as class 1 in the list of National Key Protected Animals in China. *Teinopalpus imperialis* is legally protected in India

Adults are fast and strong flyers, with males of some species being territorial. Some species can be found congregating at riverbanks, sandy areas, or hilltops. The wing coloration among species varies from pale white, yellow, and black to bright, spectacular colors, with some butterflies displaying iridescence. Many iconic species in Papilioninae have restricted distribution ranges and are dependent on specific habitats and plants.

The tribe Teinopalpini includes only the rare genus *Teinopalpus*. Two species are recognized: the Golden Kaiser-i-Hind *T. aureus* found in mountains in southern China and Vietnam and the Kaiser-i-Hind *T. imperialis*, an inhabitant of the Himalayas in Nepal and India, and also found in Vietnam. Both species are range restricted to high-quality forests in mountainous regions and these populations need conservation attention. Because of its spectacular patterns and color, this genus is often used in detailed studies on nanostructures and three-dimensional structures.

KITE SWALLOWTAILS, DRAGONTAILS, AND SWORDTAILS

This tribe includes seven distinctive genera distributed in a large variety of regions and habitats in Southeast Asia (*Lamproptera, Meandrusa*), temperate Asia (*Iphiclides*), Africa, Australia (*Graphium*), and the USA and Neotropical region (*Eurytides, Protesilaus, Protographium*, with species *P. leosthenes* in Australia). There are around 192 species and almost 500 subspecies described, with half of the entire diversity of the tribe in the genus *Graphium* spread over two continents. Adults have shorter antennae than those of other tribes, and their front legs (tibia and tarsi) are covered with scales. This group includes species with or without tails, such as Neotropical *Eurytides euryleon*.

Size varies greatly, and the White Dragontail *Lamproptera curius* can be as little as ¾ in (20 mm) in its forewing length, despite its hindwings having "tails" that are as long as its body, doubling its size.

RIGHT | The Mexican Kite-Swallowtail *Eurytides epidaus* gets its name from the kite-like shape of its wings and the tail-like extensions on the hindwings.

OPPOSITE LEFT | The White Dragontail *Lamproptera curius* is the smallest butterfly in the Papilionidae family and is found in Southeast Asia.

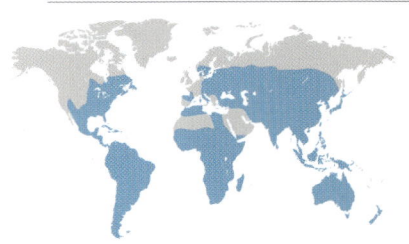

GENERA
Lamproptera, Protographium, Eurytides (including *Mimoides*), *Protesilaus, Iphiclides, Graphium*, and *Meandrusa*

DISTRIBUTION
Every continent except Antarctica

HABITATS
Widely distributed across multiple habitats, from cold tundra to deserts, in mountains with primary forest, secondary habitats, bare hills, and gardens, from lower elevations up to approximately 6,500 ft (2,000 m)

SIZE
Medium to large: ¾–2 in (20–50 mm)

HOST PLANT FAMILIES
Annonaceae, Monimiaceae, Lauraceae, Rosaceae (*Iphiclides*), Anacardiaceae, Fabaceae, Apocynaceae Atherospermataceae, Clusiaceae, Dioscoreaceae, Fabaceae, Sapindaceae, Sapotaceae, Winteraceae (*Graphium*),

Species do not show sexual dimorphism with females being very similar to males, although their colors are more muted and they are slightly larger in some species. The females of many species are unknown. Males of some species are easy to recognize because they have a fold in the hindwing, which contains long, specialized, hair-like scales. This can be seen in, for example, the colorful Purple Spotted butterfly *Graphium weskei*.

Males congregate in mud puddles in forests or along riverbanks, including African species of the genus *Graphium* and Neotropical *Eurytides*. Adults are fast flyers and some species mimic other species that are poisonous to predators.

Magnoliaceae, Euphorbiaceae, Verbenaceae (*Protesilaus*, *Graphium*), Lauraceae (*Meandrusa*), Annonaceae (*Protographium*, *Eurytides*, *Protesilaus*, *Graphium*), Hernandiaceae, and Rutaceae (*Lamproptera*, *Graphium*)

TOP | The taxonomy of the beautiful American Zebra Swallowtail *Eurytides marcellus* has changed up to three times recently, named *Eurytides*, *Neographium*, or *Protographium*.

ABOVE | An adult of the Macleay's Swallowtail *Graphium macleayanus*. This pretty butterfly has been featured on stamps in its native Australia.

BIRDWINGS, CATTLEHEARTS, CLEARWINGS, AND ARISTOLOCHIA SWALLOWTAILS

BELOW | A female of the Australian Clearwing Swallowtail *Cressida cressida*, displaying her wings lower, probably signaling that she is already mated.

This tribe includes 137 species and more than 500 subspecies arranged in nine genera. Some are inhabitants of Southeast Asia only (*Atrophaneura*), some are also in Australia (*Cressida, Troides* [including *Ornithoptera, Trogonoptera*]), and others are in the Neotropical region (*Euryades, Parides, Battus)*; the only representative in the Afrotropical region is Madagascan endemic *Pharmacophagus antenor.* Butterflies are highly variable in their shape and size, and include some of the most spectacular species—among the most sought after by naturalists and collectors. Females of the world's largest butterfly, the Queen Alexandra Birdwing *Ornithoptera alexandrae*, an endemic of Papua New Guinea, have forewings reaching lengths of up to 4 in (100 mm). Many species do not have tail streamers on their hindwings, as in males of the Australian *Cressida cressida*; however, they are distinct from females in appearance, as

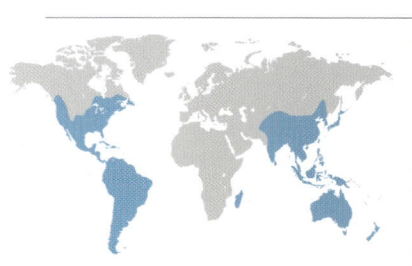

GENERA
Atrophaneura, Battus, Byasa, Cressida, Euryades, Parides, Pharmacophagus, and *Troides* (including *Ornithoptera* and *Trogonoptera*)

DISTRIBUTION
Southeast Asia, Australia, USA, Central and South America, and Madagascar

HABITATS
Tropical forests

SIZE
Large: 1½–4 in (40–100 mm)

HOST PLANT FAMILIES
Aristolochiaceae

CONSERVATION
Endemic species *Atrophaneura luchti* (Indonesia), *Atrophaneura jophon** (Sri Lanka), *Ornithoptera alexandrae* (Papua New Guinea) are categorized as Endangered by the IUCN. All species known as birdwings—*Ornithoptera,*

well as in size, with females being smaller, an unusual feature in butterflies.

In *Ornithoptera* the sexes show remarkable dimorphism, with males smaller than females and displaying metallic green. Males in *Parides* and in some species in *Atrophaneura* have characteristic unrolled pouches filled with deciduous hair-like, lower-level androconia. All species in Troidini feed exclusively on plants in the Aristolochiaceae family, which contain toxic chemicals; caterpillars successfully sequester these chemicals without harm and become unpalatable and inedible for predators. Caterpillars of some species are camouflaged by looking like bird droppings, while others have warning coloration, deterring predators. Caterpillars frequently leave the host plant to pupate elsewhere. The spectacular appearance of charismatic species such as the emblematic Queen Alexandra Birdwing *Ornithoptera alexandrae* and Wallace's Golden Birdwing *Ornithoptera croesus* has helped raise awareness of the conservation of butterflies in general, benefiting areas and species that are less charismatic to the public.

RIGHT | One of the world's most iconic species and the biggest butterfly in the world, the Queen Alexandra Birdwing *Ornithoptera alexandrae.*

Troides, and *Trogonoptera*—and *Atrophaneura pandiyana* (India) are listed in CITES, as are those marked with *.
Brazilian endemics *Parides bunichus chamissonia, P. burchellanus*, P. klagesi,* and *P. panthonus castilhoi* are listed as Critically Endangered and *P. ascanius* and *P. tros danunciae* as Endangered in the Red Book of Endangered Fauna of Brazil

FLUTED SWALLOWTAILS

The latest molecular studies favor including all species in one large genus, *Papilio*, but recognizing 14 subgenera (see panel), many of which are widely used in the literature as genera in their own right. Butterflies in this tribe have scaleless antennae. *Papilio* is the largest genus with

BELOW | A male of the African Mocker Swallowtail *Papilio dardanus*, a species widely studied by geneticists, because of the evolution of different mimetic wing patterns in the females.

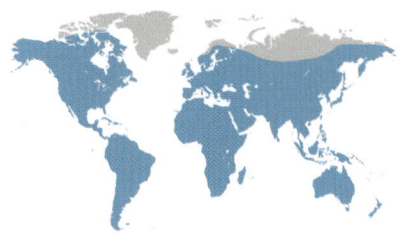

GENERA
Papilio, including subgenera *Achillides, Alexanoria, Araminta, Chilasa, Druryia, Eleppone, Euchenor, Heraclides, Menelaides, Nireopapilio, Papilio, Princeps, Pterourus,* and *Sinoprinceps,* which are often treated separately

DISTRIBUTION
Global except Antarctica

HABITATS
Widely distributed across multiple habitats

SIZE
Large: 1¾–3¼ in (45–80 mm)

HOST PLANT FAMILIES
Aristolochiaceae, Anacardiaceae, Apiaceae, Asteraceae, Araliaceae, Araceae, Betulaceae, Bignoniaceae, Brassicaceae, Campanulaceae, Canellaceae, Cannabaceae, Clusiaceae, Cornaceae, Fabaceae, Grossulariaceae, Juglandaceae, Lauraceae, Loganiaceae, Magnoliaceae, Malvaceae, Meliaceae,

Monimiaceae, Moraceae, Myrsinaceae, Nitrariaceae, Oleaceae, Paeoniaceae, Piperaceae, Poaceae, Polygalaceae, Primulaceae, Proteaceae, Rosaceae, Rhamnaceae, Rubiaceae, Rutaceae, Sapindaceae, Sapotaceae Salicaceae, Simaroubaceae, Staphyleaceae, Styracaceae, Tamaricaceae, and Ulmaceae

CONSERVATION

Species categorized as Endangered on the IUCN Red List include the country endemics *Papilio* (*Pterourus*) *homerus** (Jamaica), *Papilio* (*Pterourus*) *esperanza* (Mexico), *Papilio chikae** (Philippines), *Papilio godeffroyi* (Samoa), and *Papilio lampsacus* (Indonesia). Endemics *Papilio hospiton* (Corsica and Sardinia), *Papilio phorbanta* (Reunion Islands), are listed in CITES, as are those marked with *. Brazilian *Heraclides himeros baia* and *Heraclides himeros himeros* are listed as Endangered in the Red Book of Endangered Fauna of Brazil

TOP | The Swallowtail *Papilio machaon* is a widely distributed species and one of the largest butterflies found in Europe.

ABOVE | A female Great Mormon *Papilio memnon* resting on leaves.

up to 235 species recognized, many of which have received different taxonomic treatments at generic level. The proposed subgenera *Heraclides*, *Chilasa*, and *Pterourus* are distinct from other *Papilio* (having diverged 25–30 mya); it remains to be seen whether the single genus arrangement will be widely accepted. Females of the African Mocker Swallowtail *Papilio dardanus* form complex mimetic rings, mimicking species in Nymphalidae subfamilies such as Danainae and tribes Amathusiini and Acraeini, and also large day-flying moths, displaying an array of different wing patterns. It has been considered by some geneticists to be "the most interesting butterfly in the world."

Host plant specialization can vary dramatically. Caterpillars of the Mexican endemic *Papilio* (*Pterourus*) *esperanza* feed exclusively on the plant *Magnolia dealbata* (Magnoliaceae), which is also endemic. In contrast, caterpillars of North American *Papilio zelicaon* feed on at least 45 different plant species. Caterpillars of some species can be serious pests: for example, Lime Swallowtail *Papilio demoleus* damages lemon and orange groves and curry plants in southern Asia, Australia, and, more recently, the Caribbean where it was recently introduced. *Papilio* (*Heraclides*) *anchisiades* is also a citrus pest in South America. This tribe includes the African Giant Swallowtail *Papilio antimachus*, which gets its common name from being the largest species of butterfly on the African continent, with a wingspan of over 6 in (~155 mm).

PAPILIONIDAE: PRAEPAPILIONINAE FOSSIL BUTTERFLY

This butterfly subfamily is not often found in books because it includes only species described from fossils, which have probably been extinct for millions of years. In 1978, North American researchers discovered several important fossil butterflies in Colorado, from which they described two species, both known only from there, *Praepapilio gracilis* and *P. colorado*. The elaborate fossils were found in deposits from the Eocene period, dating from approximately 48 mya. These medium-sized species lack long tail streamers, but the hindwing has a protruding center. With fossils of this age, no coloration or scale patterns can be seen, and nothing is known of the caterpillars, eggs, host plants, or ecology. However, there are well-preserved examples in which the venation and overall shape of the butterflies can be studied in detail. The fossil butterflies share some elements of their wing structures, which are unmatched in extant swallowtail subfamilies. Based on recent cladistic and molecular studies, and considering the age of the Eocene deposits in which these

fossils were discovered, it is thought that representatives of this subfamily occurred before any modern-day Papilionidae. However, *Praepapilio* is likely to have coexisted with ancestors of *Baronia brevicornis*.

This is not the oldest known butterfly fossil—there is a Hesperiidae dating from 54 mya—but unlike *Praepapilio*, that fossil is considered closely related to a subfamily that still exists today. These ancient Papilionidae flew during the same period when many modern groups of mammals also first arose, including cetaceans, ungulates, bats, primates, and rodents.

BELOW | The fossil butterfly *Praepapilio colorado* discovered in the USA is well preserved and shows its wings spread.

HEDYLIDAE
MOTH-LIKE BUTTERFLIES

The moth-like butterflies in the Hedylidae family are only found in Central and South America, where they inhabit forests up to 10,000 ft (3,000 m) elevation. There are currently 39 recognized species, although a few more from remote areas remain to be described, as well as specimens that are in museum collections. Generic relationships among the family remain uncertain. Previously, species were separated into different genera, but then placed together in the single genus Macrosoma. However, a recent comprehensive but unpublished study by a Brazilian researcher has reinstated the two former genera Hedyle and Phellinodes with Macrosoma.

Due to their dull appearance and limited distribution, hedylids have received little taxonomic study or collection in the field. Wing patterns among males and females are almost undistinguishable from one another, although females can be paler and slightly larger—for example, in Macrosoma (Phellinodes) lucivittata. Although distributed in various countries in the Neotropical region, there is poor representation of specimens in museum collections and most published studies have focused on species from Costa Rica.

The position in the Lepidoptera of this unusual, small family of moth-like butterflies has been uncertain historically. Its classification was confusing because species in Hedylidae have typical moth traits such as antennae that are filiform, not clubbed as in most butterflies; in some species they are even bipectinate (comb-like). Also, they have a fully operating frenulo-retinacular (spine and hook) coupling system, which attaches the forewings and hindwings together.

The hedylids are mostly nocturnal and are attracted to light traps at night. They have specialized anti-bat ultrasound hearing organs. Until the 1980s, these butterflies were included in the moth family Geometridae due to their similarity in appearance to them. However, there are reports of *Macrosoma tipulata* flying during the day. Further morphological and, more recently, molecular studies, have confirmed their closer relationship, within the true butterflies, to the superfamily Papilionoidea.

GENERA
Hedyle, Macrosoma, and *Phellinodes*

DISTRIBUTION
Neotropical: Central and South America

HABITATS
Primary forest and nearby open places in tropical areas, including the Amazon Basin, from sea level to 10,000 ft (3,000 m)

SIZE
Small to medium (slim): ½–1 in (14–27 mm)

HOST PLANT FAMILIES
Tiliaceae, Sterculiaceae, Bombacaceae, Malvaceae, Melastomataceae, Euphorbiaceae, and Malpighianeae

CONSERVATION
No assessments have been made for species in Hedylidae and so no information about the conservation of species in *Macrosoma* is available. There are no Red List species. However, forests and species in South America are threatened by

Abdomens in males and female adults are slim. In many species the wings are elongated in shape and delicate, with dull brown, gray, and cream patterns. Counterintuitively, molecular studies concluded that these are not the most primitive of butterflies, a position reserved for the Papilionidae (swallowtails). Hedylidae is now considered to be the sister group of the Hesperiidae (skippers).

The biology and immature stages of most species in the Hedylidae family are little studied. However, a few detailed studies of *Macrosoma* (*Hedyle*) *semiermis* and *M. tipulata* are available. These species have elongated eggs very similar to those of pierids in shape, texture, and pale yellow color; the eggs are laid individually in young leaves. The larvae have horns on their heads, similar to some species in the brush-footed family, Nymphalidae, and can be brown. Pupae are green, flattened on one side, and are attached to the substratum by a silk cord. There are many more species whose lifecycles are unknown than known in this subfamily.

accelerating deforestation rates, which have almost tripled in the last ten years in the Brazilian Amazon, driven by conversion to agriculture for cattle pasture and crops such as soybeans. Species at high elevations are also threatened due to agriculture and, likely, changes in climate

ABOVE | One of the most common moth-like butterflies, *Macrosoma tipulata*. Caterpillars feed on leaves of the Amazonian cupuaçu *Theobroma grandiflorum* (Malvaceae).

OPPOSITE | Moth-like *Macrosoma cascaria* is found from Mexico to Costa Rica.

HESPERIIDAE
SKIPPERS, LONGTAILS, FLATS, AND AWLS

The skippers are a family of unusual, dull-colored, small- to medium-sized butterflies that have been challenging taxonomists for years. The family is one of the most diverse butterfly groups, comprising approximately 4,500 species and more than 1,300 subspecies. The family is currently divided into 13 recognized subfamilies. Skippers are found on all continents except Antarctica, at elevations from sea level up to 16,000 ft (5,000 m). They are most common in tropical latitudes, with their greatest diversity in the Neotropical region and absent from New Zealand.

Butterflies of the Hesperiidae family are very distinct in their morphology and flight patterns. They have large, stout bodies, contrasting with their short wings, and large heads often broader than the body. In most species, wings are held at an elevated v-shaped angle when perched due to this family possessing a frenulum and retinaculum, which join the forewing and hindwing. Such features are also found in moths, but are absent in all other butterflies. Many species have brown and gray coloration and somber wing patterns (some mottling, but no ocelli), although many others have bright markings, patterns, and coloration. Due to their features, for centuries skippers were regarded as a separate superfamily (Hesperioidea) from the butterflies, which were thought to be more closely related to moths.

TOP ROW | (**Left**) Colorful firetips in the genus *Mysoria* get their common name from the prominent tuft of red or orange scales at the end of the abdomen. (**Right**) A flat skipper from the *Celaenorrhinus maculosus* complex. This is a hyper-diverse group that is distributed worldwide.

In adult skippers, all three pairs of legs are developed for walking. The base of the antennae is widely separated, and the shape of the tip is curved or hooked. Their eyes are smooth with reduced ommatidia. The forewing is triangular. Adults prefer to fly in bright sunshine in the morning, although some species fly at dawn. Their rapid, bouncy, and jerky flight over short distances was the inspiration for their common name: skippers.

Skippers' eggs are dome-shaped or hemispherical and sculpted in various ways, differing in each subfamily, and are usually laid singly. Larvae are often cylindrical and smooth, although some have setae on the head or body. They have a constriction in their neck that makes the large head more prominent. Caterpillars of some species feed at night and keep out of sight during the day in shelters they make by rolling leaves and tying them together with silk. Most are dull and have no strong pigmentation, relying on their nocturnal

BOTTOM ROW | (**Left**) The spectacular skipper *Pythonides jovianus*, an inhabitant of pristine forests in Colombia. (**Right**) The White-banded Awl *Hasora taminatus* is often confused with other similar species and is found from India to Australia.

habits and camouflage as survival strategies. Pupae are long and slender and generally located in a specially spun shelter. Host plants differ markedly between subfamilies, with only one feeding on monocotyledonous plants, such as grasses, while other subfamilies use dicotyledonous plants.

Molecular studies have now unequivocally shown the previously recognized superfamily Hesperioidea to be a family of the true butterflies, Papilionoidea. The skippers are one of the more ancestral families, after Papilionidae and Hedylidae. Higher-level taxonomy within the skippers has been highly dynamic and is still being untangled, with several major molecular studies published in recent years, together with some morphological and life history studies. The number of recognized subfamilies has doubled recently, with several smaller groups now representing distinct lineages. The subdivision of these subfamilies into tribes has been even more unstable, with numerous recent proposals.

ABOVE | A male adult of the Mimic Flat *Abraximorpha davidii* found in Asia. This species has been subject to several genetic analyses but little is known about its biology.

AWLS, POLICEMEN, AND GIANT SKIPPERS

This subfamily of skippers has nine recognized genera, almost 80 species, and 200 subspecies. The Coeliadinae is one of the earliest butterfly lineages to have reached Australia, around 72 mya. The unusually long and slender ends of the labial palps help distinguish this group from other Hesperiidae and are often larger and more colorful, displaying shades of metallic blue or bright yellow, and orange marks, for example, in species of the Asian genus *Choaspes*. Adults are active in the early hours of the day, flying quickly despite having robust bodies. Species are distributed in the Old World; this family has no representatives in the Neotropical region. Males have been found hilltopping, and visiting flowers and bird droppings. Adults rest with their wings held in a v-shape above the thorax, an ancestral feature.

Females lay the white dome-shaped eggs individually, with the emerging caterpillars feeding on the soft new growth of plants. Caterpillars are brightly colored with contrasting bands and use plant leaves to build tunnels, which provide them with shelter during the day. Caterpillars of most species feed on dicotyledonous plant families, except the Orange Awlet *Burara harisa*, which feeds on the monocotyledonous ginger family Zingiberaceae. Species in some genera can be polyphagous. Pupae have a projection in the head. Although the immature stages of most Australian species are well-known, most Asian and African species remain little known. Annual mass migration has been recorded for the Australian Brown Awl *Badamia exclamationis* from Queensland to Papua New Guinea, and in African *Coeliades libeon* in East and southern Africa.

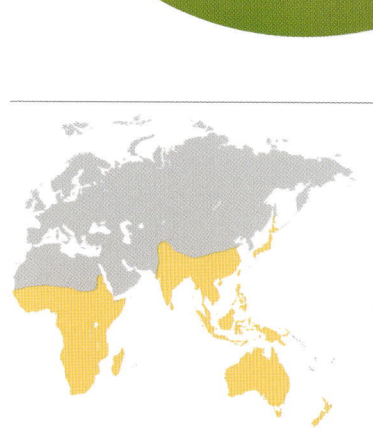

GENERA
Allora, Badamia, Choaspes, Coeliades, Pyrrhiades, Pyrrhochalcia, Hasora, Burara, and *Bibasis*

DISTRIBUTION
Indo-Australasian region; some species in Africa and the Holarctic

HABITATS
Rainforests, forest edges, and mud puddles

SIZE
Medium to large: ¾–1½ in (21–40 mm)

HOST PLANT FAMILIES
Fabaceae, Malpighiaceae, Sabiaceae, Anacardiaceae, Moraceae, Melastomataceae, Arecaceae, Rubiaceae, Euphorbiaceae, Combretaceae, Zingiberaceae, and Asclepiadaceae

OPPOSITE | The Pale Green Awlet *Burara gomata* is found in Southeast Asia. Its equally colorful caterpillars feed on leaves of the Vine *Schefflera venulosa*.

TOP | The Western Blue Policeman *Pyrrhiades lucagus* is a skipper found in West Africa. Skippers can be attracted in the field using tissue and saliva.

ABOVE LEFT | An adult of the Indian Awlking *Choaspes benjaminii*. Caterpillars of this species display bright warning colors to deter its predators.

ABOVE RIGHT | The Brown Awl *Badamia exclamationis* is a common species found in forests in India and other Asian countries. The dull coloration of some skippers makes them difficult to identify.

REGENT SKIPPER, CRESCENT SPOTTED FLAT, AND GIANT HOPPER

The subfamily Euschemoninae is monotypic, including only the spectacular Regent Skipper *Euschemon rafflesia*, of which two subspecies are recognized. This colorful species is found only in Australia, where it is restricted to rainforests. It was previously included in the subfamily Pyrginae; however, recent studies resulted in its own subfamily (proposed more than a hundred years ago) being reinstated. Males of this endemic species in Australia have a frenulum and retinaculum to couple their wings. The eggs are pale green or gray, dome-shaped, and laid individually. Caterpillars feed only on the primitive monocotyledonous family plant Monimiaceae.

RIGHT | Australian Regent Skipper *Euschemon rafflesia*, one of the largest and most colorful butterflies in Hesperiidae.

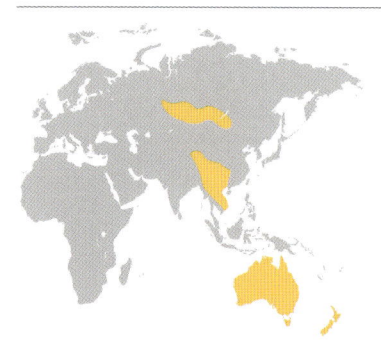

GENERA
Euschemon (Euschemoninae); *Chamunda* (Chamundinae); *Apostictopterus*, and *Barca* (Barcinae)

DISTRIBUTION
Australia (*Euschemon*), Southeast Asia (*Chamunda*), and Asia, India, and China (*Barca*)

HABITATS
Forest and forest margins

SIZE
Medium: ¾–1⅓ in (20–35 mm)

HOST PLANT FAMILIES
Monimiaceae (*Euschemon*)

ABOVE | The skipper *Barca bicolor*, an uncommon species found in north Vietnam and China, is placed in its own subfamily, Barcinae, due to its distinct phylogenetic characteristics.

The Chamundinae subfamily includes only one species, the Crescent Spotted Flat *Chamunda chamunda*. This species is found in Southeast Asia and is rarely seen. Males and females are very similar, with pale brown in the wings and a characteristic white band in the apex of the

---

CONSERVATION

The Regent Skipper *Euschemon rafflesia* is endemic to Australia and is found locally in forest patches along the east coast of the country. However, many of these habitats are now being lost to housing development and habitat fragmentation. It is recommended that detailed distribution assessment is conducted to establish its conservation status

forewings. Adults are uncommon in forests and rest with spread wings on the underside of leaves. It has been recorded at elevations from 1,600–5,000 ft (500–1,500 m). Its immature stages are unknown.

Barcinae has been described recently as a subfamily following molecular studies. This subfamily includes two monotypic genera distributed in Asia. The dull-colored Giant Hopper *Apostictopterus fuliginosus* is found in India and China and has two subspecies recognized. The skipper *Barca bicolor* is found in mid and high elevations in the Himalayas and has a characteristic bright yellow band across its forewings. Both species have been moved from other subfamilies: Heteropterinae and Hesperiinae and, more recently, Trapezitinae. Further studies are needed for these species not only to establish their position within Hesperiidae but also to discover more about their immature stages and host plants.

LONGTAILS, HAMMOCK, BROWN LONGTAIL, AND TWO-SPOTTED SCARLET-EYE

The Eudaminae is one of the most species-rich subfamilies, with approximately 600 species and 160 subspecies arranged in 53 genera and four tribes. Butterflies in this group are inhabitants of the Neotropics, with the sole exception of the Asian genus *Lobocla*. They are robust and vary greatly from small and completely brown to brightly colored, including yellow species in the genus *Entheus*, white or metallic blue in, for example, species of *Tarsoctenus* and *Phocides*, and semi-transparent wings as in *Phanus*. Some genera, such as *Urbanus* and *Chioides*, have tail-like features in the hindwings. Females have accessory glands in their genitalia, and males have androconial scales in the forewing, presumably scent organs. Dimorphism can be extreme, for example in the genus *Entheus*.

Adults are fast flyers and are active during sunny hours, although some species fly at dawn. The large diversity of species in this subfamily is reflected in the contrasting behavior and distribution of some local species in primary forests; others, such as the

LEFT | The common Ghost-Skipper *Phanus marshalli* is a species widely distributed in Central and South America. Glasswings are unusal in Hesperiidae—the effect is produced by very thin and narrow scales on the surface of the wings.

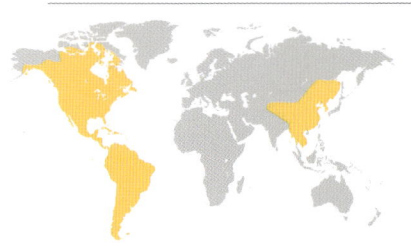

GENERA
Augiades, Drephalys, Entheus, Hyalothyrus, Phanus, Pseudonascus, Tarsoctenus, and *Udranomia* (Entheini); *Aguna, Astraptes, Autochton, Cecropterus, Cephise, Chioides, Codatractus, Ectomis, Epargyreus, Lobocla, Lobotractus, Narcosius, Polygonus, Proteides, Ridens, Spathilepia, Spicauda, Telegonus, Telemiades, Urbanus, Venada, Zestusa,* and *Zeutus* (Eudamini); *Cogia, Marela, Nerula, Oechydrus, Oileides,* and *Typhedanus* (Oileidini); *Aurina, Bungalotis,* *Dyscophellus, Emmelus, Euriphellus, Nascus, Nicephellus, Phareas, Phocides, Porphyrogenes, Salantoia, Salatis,* and *Sarmientoia* (Phocidini)

DISTRIBUTION
Mostly Neotropical; genus *Lobocla* occurs in Southeast Asia

HABITATS
Primary and secondary forests, paths, and gardens

Long-tailed Skipper *Urbanus proteus*, are widespread. Caterpillars can be a minor pest of bean plantations.

Females lay several conical eggs at a time in the terminal shoots of the host plant. Caterpillars are colorful with large heads. Pupae are dark brown with a pair of black, raised false eyespots at the wing base. Some species form a shelter from leaves, which is suspended and supported by a girdle. More studies are needed as the immature stages of many species in Eudaminae are unknown.

LEFT | *Lobocla bifasciatus* is a skipper found in Indochina. Skippers are fast flyers, making them difficult to collect.

BELOW LEFT | White-striped Longtail *Chioides albofasciatus*, a species found in North America and Mexico, is active during sunny hours and frequently visits forest edges.

ABOVE | Metallic blue skippers in the complex *Telegonus fulgerator* comprise several similar-looking species found in the Neotropical region. Species can only be separated by comparing different characteristics, for example, wing morphology, DNA, and immature stages.

SIZE
Small to medium: ⅓–1¼ in (11–30 mm)

HOST PLANT FAMILIES
Fabaceae, Urticaceae, Sapindaceae, Myristicaceae, Combretaceae, Myrtaceae, and Poaceae (*Urbanus teleus*)

CONSERVATION
Brazilian endemic species *Drephalys mourei* is categorized as Critically Endangered and *Drephalys miersi* as Endangered in the Red Book of Endangered Fauna in Brazil. No species in this subfamily has been assessed for inclusion on the IUCN Red List

FLATS, SPRITES, ELFINS, PARADISES, RAGGEDS, WHITE-CLOAKED, AND BUFF-TIPPED

Tagiadinae is very diverse and includes approximately 328 species and 267 subspecies arranged in 35 genera and three tribes; they include some genera previously included in the subfamily Pyrginae, such as the hyperdiverse and widespread genus *Celaenorrhinus*. This subfamily includes very attractive and colorful skippers and groups that can be recognized at a glance. They are found mostly in Africa, India, and East and Southeast Asia, with a few genera in Australia (*Chaetocneme*, *Exometoeca*, *Netrocoryne*) and the Americas (*Celaenorrhinus*). Some species can be prevalent and widespread across several countries,

ABOVE LEFT | Common Spotted Flat *Celaenorrhinus leucocera* is widely distributed across Asia. The hooked tip of the antennae displayed is a characteristic feature of skippers in Hesperiidae.

ABOVE RIGHT | Cream Flat *Eagris denuba*, a species found in West Africa. Males feed on flowers and also on bird droppings.

while others are restricted in their distribution, such as Madagascan endemic *Celaenorrhinus ambra* and Australian endemics *Exometoeca nycteris* and *Chaetocneme denitza*.

Adults fly rapidly and rest on top of leaves with their wings expanded. Some species fly only in the

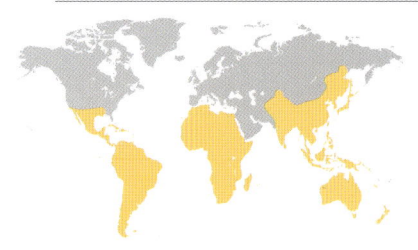

GENERA
Alenia, Apallaga, Bettonula, Celaenorrhinus, Eretis, Kobellana, Pseudocoladenia, Sarangesa, and *Scopulifera* (Celaenorrhinini); *Chaetocneme, Exometoeca,* and *Netrocoryne* (Netrocorynini); *Abantis, Abraximorpha, Calleagris, Capila, Caprona, Coladenia, Ctenoptilum, Daimio, Darpa, Eagris, Gerosis, Leucochitonea, Mooreana, Netrobalane, Odina, Odontoptilum, Pintara, Procampta,*

Satarupa, Seseria, Tagiades, Tapena, and *Triskelionia* (Tagiadini)

DISTRIBUTION
Africa, Asia, Australia, and South America

HABITATS
Rainforests, forest edges and open areas, rivers and streams, well-wooded hills, and savanna (*Netrobalane*)

SIZE
Small to medium: ½–1¼ in (12–28 mm)

mornings, while others are more active in the afternoon and dusk. Dimorphism varies among species, with some females larger than males and with wider abdomens, or with rounder wings, some with extra marks. Eggs are dark and laid individually above mature leaves. Some species hide them carefully, as recorded in the African Clouded Flat *Tagiades flesus*. Caterpillars feed on dicotyledonous plants. Many species are polyphagous; caterpillars of the Australian endemic the Bronze Flat *Netrocoryne repanda*, for instance, feed on at least 20 native plant species from different families. They also make shelters from plant leaves. Individuals of the Fritzgaertner's Flat *Celaenorrhinus fritzgaertneri* roost gregariously during the day and are active at night.

HOST PLANT FAMILIES
Acanthaceae, Sapindaceae, Anacardiaceae, Malvaceae, Sterculiaceae, Annonaceae, Euphorbiaceae, Fabaceae, Tiliaceae, Rutaceae, Dioscoreaceae (*Daimio, Tagiades*), Laureaceae, Myrtaceae, Sterculiaceae, and Cunoniaceae

CONSERVATION
Kenyan endemic skipper *Abantis meru* is categorized as Endangered on the IUCN Red List. The Black Angle *Tapena thwaitesi* is legally protected in India by the Wildlife (Protection) Act

ABOVE | The Chestnut Angle *Odontoptilum angulata* is widely distributed in India and some neighboring countries. Its name "angulata" comes from the particular shape of its forewings.

FIRETIPS

The Pyrrhopyginae have received different taxonomic treatments recently, and agreement for arrangements at tribal level have not been reached. Currently, there are six tribes recognized and around 180 species and over 160 subspecies arranged in 38 genera. Species in this group are distributed in the Neotropical region, although one species, *Apyrrothrix arizonae*, is now known to extend to the USA. Skippers in this subfamily are medium-sized and include some of the most beautiful and colorful species in Hesperiidae, with metallic and bright blues, purple, and orange in the wings, as in the Sunburst Glory *Myscelus phoronis* and *Mahotis versicolor*, one of the firetips. The common name, firetip, relates to the bright red scales at the tip of the abdomen, seen in species of *Mahotis*, *Aspitha*, and *Pyrrhopyge*, among others. Some species in genus *Jonaspyge* have red scales on the tip of the hindwings. Species in the genus *Oxynetra* have transparent wings. Mimicry complexes have been recorded in *Pyrrhopyge*, *Elbella*, and *Passova*.

LEFT | Dull Firetip *Apyrrothrix araxes* is a popular species found in the USA and Mexico. Adults are powerful flyers. Caterpillars of this species are colorful and live in oak trees.

GENERA
Azonax (Azonaxini); *Jera* (Jerini); *Oxynetra, Olafia,* and *Dis* (Oxynetrini); *Agara, Aspitha, Granila, Myscelus,* and *Passova* (Passovini); *Aesculapyge, Aerardaris, Apyrrothrix, Amenis, Amysoria, Apatiella, Ardaris, Blubella, Creonpyge, Croniades, Cyanopyge Chalypge, Elbella, Jember, Jematus, Jemasonia, Gunayan, Hegesippe, Jemadia, Jonaspyge, Mahotis, Microceris, Mimadia, Mimoniades, Merobella, Melanopyge, Mimardaris, Mysoria, Mysarbia, Nosphistia, Ochropyge Parelbella, Protelbella, Pseudocroniades, Sarbia, Sarbiena, Santea, Pyrrhopyge,* and *Yanguna,*(Pyrrhopygini); and *Zonia* (Zoniini)

DISTRIBUTION
Neotropics; one species extends to USA

HABITATS
Tropical forest, Atlantic rainforest, and along streams with primary and secondary vegetation

Adults are strong, fast flyers found in forests and puddling along streams. Caterpillars are hairy with robust bodies and large heads, only known to feed on dicotyledonous plants. Pupae are hairy, as seen in species of *Passova*. Females of many species are unknown, and further studies related to immature stages and host plants are needed. There are reports that the caterpillars of *Sarbia consigna* are consumed by Indigenous people in Bolivia and Peru.

ABOVE | The taxonomy of species, for example Andean *Pyrrhopyge papius*, is challenging due to the combination of factors such as variation in wing patterns, broad distribution, and limited specimens in collections to study.

SIZE
Medium: ¾–1½ in (21–36 mm)

HOST PLANT FAMILIES
Araliaceae, Burseraceae, Cecropiaceae, Clusiaceae, Combretaceae, Cunoniaceae, Fabaceae, Salicaceae, Lauraceae, Malvaceae, Melastomataceae, Meliaceae, Tiliaceae, Simaroubaceae, and Acanthaceae

CONSERVATION
Brazilian endemics *Pseudocroniades machaon seabrai* and *Parelbella polyzona* are categorized as Critically Endangered and Endangered, respectively, in the Red Book of Endangered Fauna in Brazil. The firetip *Mahotis baroni* was rediscovered in 2004 after 100 years of not being found in the wild. However, no conservation assessments have been undertaken for this species

TOP | Some species in the Neotropical genus *Myscelus* are now included in the genus Agara, for example Featured Skipper *Agara perissodora*. This group of butterflies includes very colorful adults and caterpillars popular among naturalists.

GRIZZLEDS, CHECKEREDS, SPREAD-WINGEDS, FLATS, SANDMEN, PARADISES, AND EYEDS

The Pyrginae is one of the largest subfamilies, with 94 genera, over 750 species, and more than 310 subspecies currently recognized. Pyrginae diverged from other subfamilies around 42 mya and included many genera now placed in other subfamilies after recent studies. Taxonomic arrangement for this group remains a work in progress. Currently four tribes are recognized.

Distribution is worldwide although most species are found in the Neotropical region. Adults are strong flyers active mostly in sunny hours, although some species prefer twilight. They rest on the underside of leaves with wings spread out flat. Females lay large, dome-shaped eggs individually. Caterpillars of most species are chunky with large heads and feed on dicotyledonous plants.

Achlyodini is a Neotropical tribe of small butterflies with mostly brown wings, some showing metallic blue and white shades, some with odd-shaped wings (e.g., in *Gindanes* and *Quadrus*). The tribe Carcharodini includes Neotropical genera. The genus *Spialia* is found in Africa along with marbled skippers from genus *Gomalia*, which are also found in Asia, and the Palearctic genera *Carcharodus* and *Muschampia*. Males in *Spialia* are

territorial, rest on the ground and low shrubs, and visit flowers and mud puddles.

Most species in the Erynnini have mottled wings with a brown background and scales in shades of white that form a characteristic pattern. Other species have plain brown wings with some metallic blue or purple scales. The Pyrgini are mostly Neotropical although checkered-skippers in *Pyrgus* are also distributed widely across the Holarctic region. The underside of their wings is characteristic, with mottled brown and white markings resembling a checkerboard, hence their common name.

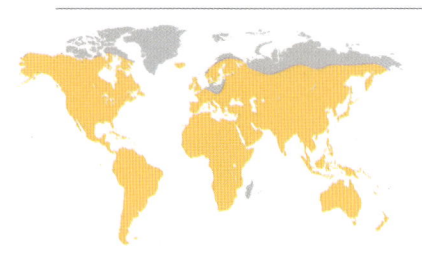

GENERA

Achlyodes, Aethilla, Atarnes, Cabirus, Charidia, Doberes, Eantis, Eboruncus, Eracon, Gindanes, Grais, Haemactis, Milanion, Morvina, Myrinia, Ouleus, Paramimus, Pseudodrephalys, Pythonides, Quadrus, Spioniades, and *Zera* (Achlyodini); *Arteurotia, Austinus, Bolla, Burca, Carcharodus, Conognathus, Cornuphallus, Cyclosemia, Ernsta, Gomalia, Gorgopas, Hesperopsis, Mictris, Muschampia, Nisoniades, Noctuana,*

Ocella, Pellicia, Pholisora, Polyctor, Sloperia, Sophista, Spialia, Staphylus, Tiana, Viola, Viuria, and *Windia* (Carcharodini); *Anastrus, Camptopleura, Chiomara, Chiothion, Clito, Cycloglypha, Ebrietas, Ephyriades, Erynnis, Festivia, Gesta, Gorgythion, Helias, Mylon, Potamanaxas, Sostrata, Theagenes,* and *Timochares* (Erynnini); *Anisochoria, Antigonus, Burnsuis, Carrhenes, Celotes, Diaeus, Heliopetes, Heliopyrgus, Onenses, Paches, Plumbago, Pyrgus, Systasea,*

ABOVE | The Orcus Checkered-Skipper *Burnsius orcus* belongs to a species complex recently allocated to a new genus after molecular studies. Little is known about the biology of this species found in the Neotropical region.

TOP LEFT | The species *Gindanes brebisson* is widely distributed from Mexico to Bolivia; however, its geographical populations are distinctive. There are four subspecies described with more to be described.

MIDDLE LEFT | Common Blue-Skipper *Quadrus cerialis*, a Neotropical species found in forests up to 4,900 ft (1,500 m).

BOTTOM LEFT | Red Underwing Skipper *Spialia sertorius* is found in Europe. This charismatic species is found in wet patches in the ground. In 2016, a detailed study separated various species in the "*sertorious*" complex.

OPPOSITE | The commonly known Glorious Blue-Skipper *Paches loxus* is a popular species because of its pretty appearance and bright blue metallic coloration, an unusal feature among the Pyrginae subfamily. The species is distributed from Mexico to Brazil, with two subspecies recognized.

Timochreon, Trina, Xenophanes, Zopyrion, and *Loxolexis* (Pyrgini)

DISTRIBUTION
Worldwide, except Antarctica

HABITATS
Temperate woodlands, alpine grassland, tropical forests, open habitats, and savanna (African *Sarangesa phidyle* only)

SIZE
Small: ½–¾ in (13–18 mm)

HOST PLANT FAMILIES
Malvaceae, Convolvulaceae, Euphorbiaceae, Plantaginaceae, Polygonaceae, Rosaceae, Sterculiaceae, and Tiliaceae

CONSERVATION
No species in this subfamily have been categorized as threatened on the IUCN Red List. The Cinquefoil Skipper *Pyrgus cirsii* is categorized as Vulnerable on the European Red List of Butterflies.

Range-restricted and endemic species, such as the Colombian endemic *Nisoniades suprapanama* in Pyrginae deserve detailed conservation assessments to establish whether they are threatened

SCARCE SPRITES

These two small Afrotropical subfamilies were described in 2022. Subfamily Katreinae was described based on molecular studies, though previous morphological revisions suggested the separation of the genus *Katreus* from their relatives in the subfamily Pyrginae. The five species previously recorded in the genus *Katreus* are now separated into two genera, *Katreus* (one species) and *Ortholexis* (four species). Butterflies in Katreinae are found in undisturbed wet habitats in West Africa. The Giant Scarce Sprite *Katreus johnstonii* is one of the largest African skippers. Adults have been reported patrolling territory and flying in loops in morning sunshine in pristine wet forest, not higher than 3 ft (1 m) from the ground. Adult males rest on the underside of leaves. These uncommon medium-sized butterflies are dark brown with an orange band on the forewings. Males and females are similar. Immature stages are unknown.

The subfamily Malazinae was described recently, in 2020. It is monotypic, with only the genus *Malaza*, which has three known species:

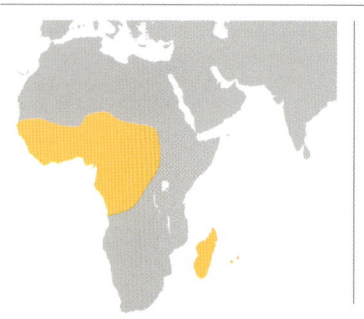

GENERA
Katreus and *Ortholexis* (Katreinae); and *Malaza* (Malazinae)

DISTRIBUTION
West Africa (*Katreus, Ortholexis*) and Madagascar (*Malaza*)

HABITATS
Undisturbed, primary, lowland forest

SIZE
Medium: 1¼ –1½ in (30–32 mm) (*Katreus, Ortholexis*); small: ½–1 in (12.5–24 mm) (*Malaza*)

HOST PLANT FAMILIES
Apocynaceae (*Ortholexis*)

M. carmides, *M. empyreus*, and *M. fastuosus*. The species in the subfamily are endemic to mainland Madagascar and include some of the most attractive skippers in Africa. This genus was previously classified in the Hesperiinae subfamily, and is an evolutionarily distinct ancient lineage among the butterflies, displaying unusual features. Species in this genus are uncommon in the wild and in museum collections. There are likely to be more unnamed species. Immature stages are almost unknown.

CONSERVATION
No species in these subfamilies have been assessed using the IUCN criteria. *Malaza fastuosus*, endemic to Madagascar, has not been collected for nearly 50 years, and it is therefore urgent that detailed assessments to establish its conservation status are undertaken

TOP | Rare adult of the Lavish Malaza *Malaza fastuosus*, a species of skipper missing from records for 50 years. The specimen was carefully bred by local photographer Armin Dett.

OPPOSITE | Giant Scarce Sprite *Katreus johnstonii*, a species found in Africa.

ABOVE | Immature stages of a rare skipper *Malaza*, photographed for the first time ever in Madagascar.

SYLPHS, CHEQUEREDS, ARCTIC SKIPPERLINGS, GOLD-SPOTS, AND SILVER-STRIPPED

This subfamily of small butterflies is arranged in two tribes and 12 genera, and has vast diversity; almost 200 species and nearly 100 subspecies have been described. The Heteropterinae displays an unusual distribution, and includes widespread genera as well as range-restricted and endemic species. Close to three-quarters of the species in the subfamily occur in South America, including the diverse gold-spots *Dalla* and the recently described *Ladda*. Other genera are found in almost all other continents—

for instance, the sylphs *Metisella* and Madagascar endemic *Hovala* in Africa, the Silver-stripped Skipper *Leptalina unicolor* in East Asia, and widespread chequered skippers *Carterocephalus* in the Holarctic region. This subfamily does not occur in North Africa, Antarctica, or Australia.

In comparison with other Hesperiidae, adults have slender bodies and are broad winged. Caterpillars feed only on monocotyledonous plants. The small size and dull brown ground color in the wings of many species make it difficult to identify

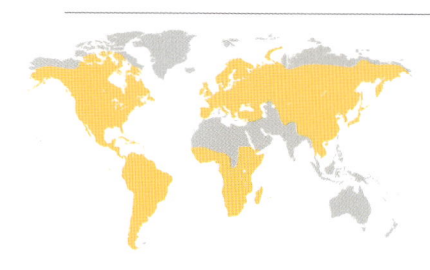

GENERA
Argopteron and *Butleria* (Butleriini); *Carterocephalus, Dalla, Dardarina, Freemaniana, Heteropterus, Hovala, Ladda, Leptalina, Metisella,* and *Piruna* (Heteropterini)

DISTRIBUTION
Worldwide, except Australasia, North Africa, and Antarctica

HABITATS
Tropical and temperate forests, damp grass,

high elevations in tropical Andes from 3,200–11,000 ft (1,000–3,400 m)

SIZE
Small: ½–⅔ in (12–15 mm)

HOST PLANT FAMILIES
Poaceae

CONSERVATION
The European Chequered Skipper *Carterocephalus palaemon* went extinct from England almost 50 years ago due to the loss of its damp grassy habitats.

species using only morphological characters. In contrast, some other species have beautiful contrasting patterns, with orange and white marks and spots, and also some with metallic coloration on the undersides of the wings, which is not as common in butterflies as in other insects. For instance, Chilean endemic *Argopteron aureum* and Colombian endemic *Dalla semiargentea* have been named because of the respective gold and silver appearance of their wings. The use of modern DNA techniques has highlighted complementing characters that bring light to the higher-level taxonomy of this group and certainly will assist with the description of many species yet to be named in this diverse group.

BELOW | Many-spotted Skipperling *Piruna aea* is a charismatic and small butterfly found from Mexico across Central America. Males rest with their wings closed, but on sunny spots they bask with their wings open wide.

However, populations have been reintroduced from continental Europe, and Scottish populations are protected under the Nature Conservation (Scotland) Act. The Asian Silver-stripped Skipper *Leptalina unicolor* is thought to be endangered because of the climate change vastly reducing its, and its host plant's, habitats. Only a few, many endemic, tropical species in this group have been subject to detailed conservation assessments

OPPOSITE LEFT | The Large Chequered Skipper *Heteropterus morpheus*, which is also known as "The Mirror" because of the patterns on the underside of its wings that resemble drops of water.

OPPOSITE RIGHT | *Carterocephalus palaemon*, widespread across the Holarctic, has many common names including the Arctic Skipperling and Chequered Skipper, the latter because the pattern on its wings resemble a checkers or chessboard.

ABOVE LEFT | Species in the genus *Dalla* are colorful, small skippers. Detailed molecular studies have recently separated many species into the new genus *Ladda*.

ABOVE RIGHT | Gold-spotted Sylph *Metisella metis*. This large genus of grasslands butterflies are widely distributed across Africa and have several species very similar to each other.

OCHRES, GRASSES, AND SEDGE-SKIPPERS

RIGHT | The Splendid Ochre *Trapezites symmomus* is commonly found in bushes of its host plant *Lomandra* in Australia. Observing groups there have reported a decline in their populations.

OPPOSITE | Spectacular view of the Australian Dingy Shield-Skipper *Toxidia peron*. The position shown, displaying the hindwings flat with the forewings elevated and angled upward, is typical of many Hesperiids when resting or perching.

This relatively small subfamily includes around 75 species and two subspecies, organized into 18 genera. They are restricted to the Australasian region with most genera and species being endemic to Australia, and some occurring in New Guinea. Butterflies in this group are small and have wings covered in dark brown scales, with white, yellow, and orange marks on the upperside in most species. The shape of the club of the antenna is a useful character for separating subfamilies. Adults have stout bodies and display rapid, erratic flight during the sunny hours. Males are territorial and display patches of

GENERA
Anisynta, Antipodia, Croitana, Dispar, Felicena, Herimosa, Hesperilla, Hewitsoniella, Mesodina, Motasingha, Neohesperilla, Oreisplanus, Pasma, Prada, Proeidosa, Rachelia, Signeta, Toxidia, and *Trapezites*

DISTRIBUTION
Australia and New Guinea

HABITATS
All major biomes in Australia (arid, alpine, coastal mountains, heathlands) from sea level to elevations up to 5,200 ft (1,600 m) in forest, open forest, subalpine woodlands, and eucalypt forests. *Rachelia extrusus* and the Spotless Grass-Skipper *Toxidia inornatus* are found in tropical rainforests.

SIZE
Small: ½–¾ in (13.5–18 mm)

short, specialized androconial scales in the forewings. There is no marked sexual dimorphism although females are rounder and larger and the spots on the forewings are more defined.

Males (but not usually genus *Toxidia*) can display hilltopping behavior. Females lay eggs individually on host plants, although in some species, such as the Wedge Grass-Skipper *Anisynta sphenosema*, females fly close to the ground and lay eggs in leaf-litter. Females are less frequently encountered in the field and museum collections.

Eggs are rounded and adorned with longitudinal ribs. Caterpillars feed on monocotyledonous plants. Some species have developing mechanisms to reduce the caterpillar's metabolic activity and development to deal with seasonal limitations of resources. Caterpillars build a shelter using leaves and stems. They rest during the day and feed at night. Pupae vary in color among groups but are often covered with a waxy powder.

HOST PLANT FAMILIES
Poaceae, Cyperaceae, Cariceae, Iridaceae, and Xanthorrhoeaceae

CONSERVATION
In recent years, populations of the Yellow Ochre *Trapezites lutea* have shown a sharp decline in some areas in Australia due to urban development. The Speckled Ochre *Trapezites atkinsi* is restricted to a small coastal headland and considered vulnerable unless further populations are discovered

DARTS, NIGHTFIGHTERS, RANGERS, PYGMIES, HOTTENTOTS, SNOW HORNEDS, BANDEDS

The Hesperiinae is the largest and most diverse subfamily of skippers, and probably the most challenging for identification in the field and in collections. Identifying species is not a straightforward process using wing characters, but might require dissecting their genitalia or using tools like molecular sampling. The group has recently been arranged into 14 tribes, with several genera difficult to place. There are over 330 genera and more than 2,100 species currently recognized. The subfamily is widely distributed worldwide with half the known species found in the Neotropics.

Adults are robust and fly swiftly and erratically in sunny areas, although some species fly at dusk or in darkness, including the African Narrow-banded Red-eye *Pteroteinon concaenira*. These small butterflies

LEFT | Hopper *Platylesches neba*, a butterfly found in southwest Africa and active at sunny times during the day. The peculiar position of its wings at rest is believed to give speed at sudden flight if disturbed.

OPPOSITE TOP | The Grass Demon *Udaspes folus* is widespread in India and Southeast Asia. Its caterpillars feed on plants of the genus Curcuma, or turmeric, widely used for its medicinal and cosmetic properties.

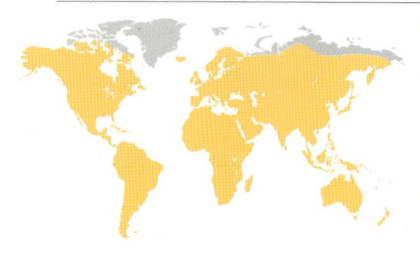

GENERA
Over 350 genera including: *Baracus* (Aeromachini); *Mopala* (Astictopterini); *Ankola* (Ceratrichiini); *Erionota* (Erionotini); *Thespieus* (Hesperiini); *Psolos* (Noctocryptini); *Telicota* (Taractrocerini); *Aegiale* (Megathymini); Caligulana (Hesperiini); *Borbo* (Baorini); *Ceratrichia* (Ceratrichiini); and Gretna (Gretnini)

DISTRIBUTION
Worldwide except Antarctica

HABITATS
Almost every known habitat, but mostly grassland; also rainforest, arid zones, woodlands, eucalypt forests, open areas, rocky areas, suburban gardens, and coastal lowlands

SIZE
Small to medium: ⅓–1¼ in (10–31 mm)

HOST PLANT FAMILIES
Poaceae (Aeromachini); Anacardiaceae, Chrysobalanaceae, Costaceae, Combretaceae, Connaraceae, Dracaenaceae, Euphorbiaceae, Fabaceae, Malpighiaceae, Liliaceae, Sapindaceae, Zingiberaceae (Astictopterini); Poaceae (Baorini, Ceratrichiini); Pandanaceae (Erionotini); Arecaceae, and Marantaceae (Astictopterini and Gretnini)

CONSERVATION
South African endemic *Kedestes sarahae* and the North American Powesheik Skipperling *Oarisma powesheik* are categorized as Critically Endangered, and North American species Arogos Skipper *Atrytone arogos*, Dakota Skipper *Hesperia dacotae*, and Ottoe Skipper *Hesperia ottoe*, and Australian Black Grass-dart *Ocybadistes knightorum* are categorized as Endangered on the IUCN Red List

ABOVE | The Chequered Lancer *Plastingia naga* is a small skipper distributed in Asia in lowland forests. The distinctive black veins, patterns, and color on the underside of the wings help to separate this species from similar ones in its genus.

keep their forewings half opened and hindwings spread flat when basking on leaves. Males feed on flowers but obtain important minerals from bird droppings. Some species of Brazilian genus *Calpodes* are known for having a very long proboscis that extends beyond the length of the abdomen. Dimorphism varies across groups. Males may display sexual secondary characters, including specialized scales in the forewings, which can be observed in North American skippers *Euphyes* and *Ochlodes*, for example. The mass migration of the African skipper *Andronymus gander* has been recorded in Cameroon and Nigeria.

The eggs of species in Hesperiinae are smooth and are laid individually. Most caterpillars feed primarily on monocotyledonous plants, although caterpillars in the Astictopterini and Gretnini tribes feed on dicotyledons. The immature stages of most tropical species are poorly documented.

LEFT | The Purple and Gold Flitter *Zographetus satwa* is found in South and Southeast Asia. Males are found visiting areas near streams and puddles.

BELOW | A male of the Large Skipper *Ochlodes sylvanus*, a widely distributed species from Europe to eastern Russia and Japan. Males can be differentiated from females by the presence of a black patch of specialized scales on the forewings.

GIANT-SKIPPERS

Butterflies in this group used to be classified in their own family, and then as a subfamily until recently. However, it is now regarded as a tribe that belongs in the Hesperiinae subfamily. The tribe gets some of its common names because it includes the largest and most robust-bodied species. They are arranged in six genera with almost 60 species and over 40 subspecies.

Species in the Megathymini are primarily found in Mexico and the USA, with only the genus *Carystoides* distributed south into Central and South America. Adults have stout, dark bodies and brown wings decorated with orange or white patterns. They have a fast and powerful flight. Although similar, females are generally larger than males and have wider and more rounded wings with additional spots. Because of their appearance, specimens of giant-skippers can be confused with moths in collections.

Females lay a large number of brown eggs, some species individually and others in clusters near the base of the host plant. Caterpillars (and adults) have narrowed heads that allow them to bore into plant stems. Contrary to popular belief, an international team of scientists found very recently, by studying DNA, that caterpillars added to bottles of tequila (a Mexican variety of mescal) are not from the Tequila Giant-Skipper *Aegiale hesperiaris* but are instead caterpillars of *Comadia redtenbacheri*, a moth from the family Cossidae. Commercial distillers use caterpillars of other species because of the high price and low abundance of the Tequila Giant-Skipper, which has likely been affected by over-collection and habitat loss. Both caterpillars are pests of *Agave* crops.

RIGHT | Arizona Giant-Skipper *Agathymus aryxna* is found in Mexico and southeastern Arizona in the USA. Adult specimens are found in open grassy woodlands.

GENERA
Agathymus, Carystoides, Megathymus, Stallingsia, Aegiale, and *Turnerina*

DISTRIBUTION
USA and Mexico; Neotropical region (genus *Carystoides* only)

HABITATS
Dry woodland, deserts, dunes, coastal areas, agave plantations, and scrub

SIZE
Medium to large: ¾–1¾ in (21–46 mm)

HOST PLANT FAMILIES
Asparagaceae (previously known as Agavaceae)

CONSERVATION
No species in this subfamily have been assessed under IUCN Red List criteria. As populations of the Mexican endemic *Aegiale hesperiaris* are in decline, a detailed conservation assessment is needed

PIERIDAE
WHITES, YELLOWS, SULPHURS, AND GHOSTS

One of the most interesting and probably understudied butterfly groups, this family comprises over 1,100 species and around 2,747 subspecies, arranged in 84 genera and four subfamilies: Pierinae, Coliadinae, Dismorphiinae, and Pseudopontiinae. It includes inhabitants of almost every habitat in the world, from sea level to the snow line of the highest mountain peaks, with the diversity of species peaking in tropical regions. Some butterflies in this family are often found in urban settings such as gardens and parks all around the world, where they can be quite tame.

Species can be small, medium, or large in size. Their yellow or white coloration is made from derivatives of uric acid. Although many think of this group as being plain in coloration, the diversity of patterns is vast, with tropical or forest-based species displaying orange-streaked tiger patterns, black coloration, or striking color combinations; some species, such as the ghosts, are even transparent. Pierids vary in size and robustness, from tiny, delicate flower-feeding butterflies to large, powerful fast flyers and long-distance migrants. Males and females often differ from one another considerably, displaying sexual dimorphism, with females often larger and paler in many species. Males often congregate at riverbanks to drink diluted salts or urine from mammals.

Adults can be distinguished from other butterfly families because of the fully developed first pair of legs, which bear bifid claws in males and females. In some other butterfly groups, the front legs are often reduced or vestigial, but Pieridae are one of several families whose butterflies use all six of their legs. In addition, the arrangement of veins in the forewings differs from those of other families.

Caterpillars are elongated and lack protuberances, although some have spines. Some caterpillars are cryptic while others patterned. There are no universal characters that can be used to describe a typical caterpillar in this family. Pupae are often elongated and hang from the cremaster, although some species develop a silk belt to keep them upright on the plant. Eggs are elongated like a rugby ball, with ribbed surfaces. Despite the diversity and abundance of species in this family, our knowledge of immature stages is scant. There are probably more species whose lifecycle is unknown than are known.

ABOVE | Several species of yellow butterflies congregating and mud-puddling. This behavior is observed more frequently in males. They are obtaining nutrients needed for reproduction directly from the soil.

Species in Pieridae are study models for biochemistry, migration, seasonal variation, and UV reflectance, although research has focused on Palearctic species; there are big gaps in our knowledge of tropical species. The latest studies on Pieridae suggest that butterflies in this family are very sensitive to local climate and particularly temperature. Many species have morphological adaptations to regulate the temperature in their environment and some have succeeded in colonizing habitats with extreme conditions.

Pierids have inspired works of art and literature; for example, a group of yellow butterflies famously accompanies one of the characters in the classic book *One Hundred Years of Solitude* by Gabriel García Márquez, winner of the Nobel Prize for Literature.

ABOVE | The taxonomic status of the Scarlet Tip *Colotis danae anae* as a recognized species or subspecies varies among specialists. The seasonal forms displayed by the *Colotis* add to the taxonomic complexity in the genus.

MIMIC SULPHURS

BELOW | *Lieinix nemesis* is characterized by its peculiar elongated and falcate forewings with metallic scales as shown in this male. Females have a similar shape but different coloration. This species is found from Mexico to Peru.

This is a very distinctive subfamily among the Pieridae because of its unusual narrow wing shape, as well as the range of colors that these butterflies display. Dismorphiinae (from the Greek meaning "abnormal shape"), includes many species with oddly narrow or oblong wing shapes, with pointed forewings and broad hindwings—for example, the Neotropical Frosted Mimic-White *Lieinix nemesis*. Most of this subfamily's 62 species and 200 subspecies, arranged in seven genera, are inhabitants of forests in Central and South America, with only one genus, *Leptidea*, found in temperate Asia and Europe. Because of their scarcity in nature and relatively scant representation in museum collections, it is likely that there are several species yet to be named.

Dismorphiinae includes small, slow-flying butterflies with white or yellow coloration, such as *Pseudopieris nehemia*. Bright coloration, including orange, black, and red scales or tiger-like patterns can also

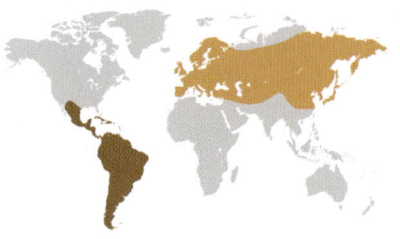

GENERA
Leptidea, Pseudopieris, Enantia, Lieinix, Dismorphia, Patia, and *Moschoneura*

DISTRIBUTION
Dismorphini (dark brown) are distributed in the Neotropical region, in Central and South America; species in the tribe Leptideini (brown) are found in the west and east Palearctic, in Europe, the Middle East, and temperate Asia

HABITATS
Species in the Dismorphiinae subfamily inhabit diverse ecosystems, including foothills to premontane and montane forest, chaparral, coniferous woodlands, and oak forests. The only genus outside the Neotropical region, *Leptidea*, occurs at alpine meadows, woodlands, roadsides, and moors. *Dismorphia* can perhaps best be found in cloud forests near streams, in mid-elevation tropical forests in South America

SIZE
Medium: ¾–1½ in (20–38 mm)

HOST PLANT FAMILIES
Fabaceae and Leguminosae

CONSERVATION
Despite the high diversity of pierid butterflies in the Neotropics, the majority of species assessed by the Red List programs are from temperate regions. The European Fenton's Wood White *Leptidea morsei* is categorized as Near Threatened on the IUCN Red List but Endangered in the European Union list EU27. The last assessment in the UK of the European Wood White *Leptidea sinapis* showed a sharp decline in populations of over 30 percent in a period of ten years. As no Neotropical species have been subject to a comprehensive assessment of the threat status, assessments in this subfamily are urgently needed, as tropical forests support the greatest diversity in this group and are known to face increasing rates of deforestation

ABOVE | The Wood White *Leptidea sinapis* has become a study model for speciation since 2011, when a complex of cryptic species was found; what was once thought to be a single species was reclassified into three distinct species.

be seen in this group—for example, in *Patia cordillera*. Antennae are clubbed but have a scaleless terminal area where specialized sensory organs called *sensilla* are sometimes clustered. The five radial veins in the forewings usually all arise from a common stalk, considered an important character to distinguish the subfamily. The wing veins are covered with scales and can be hard to see.

Males and females in the genus *Dismorphia* can show strong dimorphism, displaying very different wing characters and colors from each another. Size varies from tiny, slim, and delicate butterflies, such as in the genus *Moschoneura*, to large, strong flyers, such as in the genera *Patia* and *Dismorphia*.

The immature stages, particularly eggs, in the Neotropical genera have, unusually, been subject to detailed studies, mainly by Mexican researchers. Eggs are elongated, resembling cacao seeds of microscopic size, $1/32$–$1/16$ in (1–2 mm). Contrary to

historical claims, the color of the eggs is determined by the contents of the chorion and can vary if fertilized and as the embryo grows inside.

Caterpillars in *Dismorphia* are elongated and later instars are green in color. However, little is known about the full lifecycle of the majority of species in the Dismorphiinae.

OPPOSITE | Butterflies in the genus *Moschoneura* imitate the appearance of unpalatable glasswing butterflies to trick predators.

BELOW | The colorful *Dismorphia spio* is found in various Caribbean islands including Puerto Rico, Hispaniola, and the Dominican Republic.

YELLOWS AND SULPHURS

BELOW | Clouded Yellow or Sulphur butterfly, in the genus *Colias*. Some species display marked dimorphism. Females look distinctive and can be difficult to match with the right male.

The Coliadinae are mostly medium- to large-sized butterflies, although a few, such as the Dainty Sulphur *Nathalis iole*, are very small. They have yellow or white coloration in the wings. This cosmopolitan subfamily has the highest diversity of species concentrated in the Neotropics of Central and South America. It includes 17

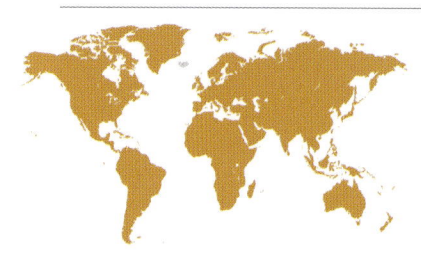

GENERA
Colias, Abaeis, Teriocolias, Zerene, Phoebis (now includes *Rhabdodryas*), *Aphrissa, Anteos, Catopsilia, Gonepteryx, Dercas,* and *Gandaca* (Coliadini). *Prestonia, Kricogonia, Nathalis, Eurema, Pyrisitia,* and *Leucidia* (Euremini)

DISTRIBUTION
Worldwide, except Antarctica and Iceland

HABITATS
Temperate species are found in open habitats and in the countryside, including coastal cliffs. In North America, temperate species are also found at high to very high elevations and very high latitudes. Neotropical species are found in dry forests at sea level, forests at mid and very high elevations (*Nathalis*, above 9,800 ft/ 3,000 m), secondary forests, and open areas. In Africa, *Colias* are found in temperate and montane grasslands up to

genera, approximately 208 species, and over 500 subspecies, arranged in two tribes, Coliadini and Euremini. This group includes the genus *Colias* (known as clouded yellows in the Palearctic region and sulphurs in the Neotropics), which is distributed worldwide and which is among the most species-rich of all butterfly genera, with around 85 species and 120 subspecies recognized. Other genera, such as *Gonepteryx*, are found only in Europe and temperate Asia, *Catopsilia* in Australia, India, and Indonesia, and *Dercas* in Southeast Asia; others are concentrated in tropical habitats in the Neotropical region, such as *Pyrisitia*, *Kricogonia*, and the distinctive Mexican endemic *Prestonia clarki*, the only species in its genus.

In most species, the appearance of males and females is similar, and not as markedly distinct as in some other butterfly groups, but on occasions males can be yellower while females are whiter or vice

ABOVE | The Brimstone butterfly *Gonepteryx rhamni* can be found in almost any habitat in the Palearctic region from Europe to Asia. Buckthorn plants are the host plants of its caterpillar.

8,900 ft (2,700 m) elevation and *Eurema* up to 8,200 ft (2,500 m) in grasslands, forest edges, coastal bush, lowland forests, marshy areas in forest, and moist savanna

SIZE
Medium: ¾–1 ⅓ in (20–35 mm)

HOST PLANT FAMILIES
Plants from Fabaceae (for the genus *Colias*, *Catopsilia*, *Zerene*, *Phoebis*, *Eurema*), Ericaceae (*Colias*), Rhamnaceae (*Gonepteryx*), Connaraceae (*Gandaca*),

Asteraceae (*Nathalis*), Salicaceae, Sapindaceae, Malvaceae, Thymelaeaceae, Oxalidaceae, Gentianaceae, and Araceae

CONSERVATION
Only a few species in this subfamily have formally been assessed for threat using IUCN criteria. *Gonepteryx cleobule* is categorized as Vulnerable, *Colias phicomone* is categorized as Near Threatened on the European Union EU27 list. Less than 1 percent of the species in the subfamily

have been assessed. Some species need attention, such as *Colias myrmidone*, which might now be extinct in Austria as it has not been seen there for more than 25 years. *Prestonia clarki* is endemic to Mexico and little is known about its conservation status

versa. The males of various species have iridescence due to nanostructures in the scales covering the wings and have been subject to fascinating studies on ultraviolet reflection—for example, *Colias eurytheme*, *Colias electo*, *Eurema hecabe*, and *Phoebis argante*. Butterflies in the Coliadinae have the humeral vein greatly reduced or absent in the hindwing. Caterpillars, which are described for some species, are mostly slim and green looking. Pupae are also usually green, but in some species brown or dark yellow.

Although pretty, some caterpillars are pests: *Colias eurytheme*, for example, can voraciously attack entire alfalfa fields (Fabaceae) in Mexico and North America. Some other iconic taxa in this group include the "dog face" butterflies in the genus *Zerene*, whose vernacular name relates to the unusual patterns in their wings. The richly described diversity of pierids includes the

ABOVE | The Lemon Emigrant *Catopsilia pomona* is widely spread thoughout India and Southeast Asia to northern Australia. Its common name is given after migratory events recorded in India, Thailand, and Australia.

infamous hoax butterfly "*Papilio ecclipsis*," which even tricked the father of taxonomy, Linnaeus, who described the supposed species with that name. It was later found that this was a painted or dyed specimen of the common Brimstone *Gonepteryx rhamni*, a widespread European species.

LEFT | *Dercas verhuelli*, commonly known as the Tailed Sulphur, is found in Southeast Asia. Little is known about the biology of some species in this genus.

BELOW | Males of different species of pierids assemble at riversides and ground wet areas, puddles, and mud to extract salts and nutrients necessary for mating. This behavior is called puddling and is rarely observed.

GHOSTS

The ghosts live in African rainforests and get their common name from their small body and wings that have no visible markings and are white, almost transparent. Their peculiar wing venation and antennae are remarkably distinct from other butterflies in the Pieridae family. The first species described in the genus was given the specific epithet "*paradoxa*" because of the contradictory character of its antennae not being clubbed as in all other butterflies in the Pieridae family. The name of the subfamily

Pseudopontiinae means "like petals," reflecting their delicate nature. Until only a decade ago it was thought that the only genus in the subfamily *Pseudopontia* had only one species, with two subspecies. These were defined due to their appearance and marked distribution in two separate areas south and north of the equator in Africa. However, after recent molecular phylogenetic studies, there are now considered to be five recognized species: *Pseudopontia paradoxa, P. australis, P. mabira, P. gola,* and *P. zambezi.*

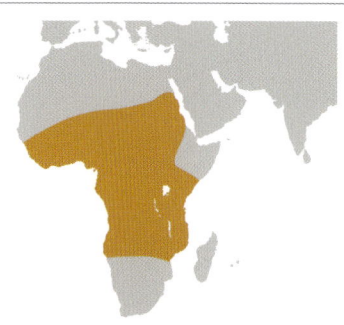

GENERA
Pseudopontia

DISTRIBUTION
Afrotropical region, mostly through West Africa

HABITATS
Near the canopy in dense, wet tropical forests and across an expanse of sub-Saharan Africa

SIZE
Medium: ¾–1 in (22–28 mm)

HOST PLANT FAMILIES
Acanthaceae and Opiliaceae

CONSERVATION
No species in the genus *Pseudopontia* has been evaluated using IUCN Red List criteria. However, *P. paradoxa* has been assessed as nationally Vulnerable in Uganda. Africa had the largest area of wooded land in the world; however, in recent decades its tropical forests have faced a fast rate of conversion of land due to expansion of

Males in *Pseudopontia* have been observed hovering above plants in groups and flying erratically, slowly and unevenly, under the canopy. These butterflies have bright green eyes.

As in many tropical species, the natural history of these butterflies is little known, and its host plant was not determined until 200 years after the species' discovery. Although the wings are almost colorless to the human eye, the scales of the forewings of male *Pseudopontia australis* reflect ultraviolet light, although those of females do not. The pupal features, wing venation, and antennae in Pseudopontiinae are considered a combination of primitive but specialized stages.

transport and communications networks, human settlements, agriculture, and other uses. The weak reporting of forest-sector statistics on the continent does not permit the real impact of those practices to be monitored

ABOVE | The Ghost butterfly *Pseudopontia paradoxa*. It was recently discovered that males of the subspecies *australis* have UV-reflecting scales, whereas males of the subspecies *paradoxa* do not.

OPPOSITE | Although uncommon in scientific collections, museum specimens of the Ghost butterfly *Pseudopontia paradoxa* have become useful to distinguish species that look alike or cryptic in the wild but, with detailed studies, can be separated.

WHITES

This is the largest subfamily of Pieridae and is vastly diverse with almost 900 species and over 2,700 subspecies. The greatest diversity is reached in tropical regions, but the subfamily is distributed worldwide, except in Antarctica. Species are found from sea level, such as the Palearctic species *Pieris napi*, up to 13,000 ft (4,000 m) high elevation peaks in the mountains of the Andes, including Colombian endemic *Reliquia santamarta*. There are about 60 genera recognized, arranged in seven tribes; however, numbers vary according to different studies and authors. Pierinae includes some of the most spectacular butterfly species, including the speciose and striking genus *Delias*, known as Jezebels, the world's richest with 251 species and 350 subspecies. Among the characters to recognize in the subfamily are their well-developed palpi and long antennae,

BELOW | *Catasticta flisa*. A recent molecular study suggested that the more than 100 species in this genus should be reassigned to the genus *Archonias*. However, due to the limited evidence provided, *Catasticta* continues to be widely accepted by most workers in the field.

GENERA
Aoa, Aporia, Appias, Archonias, and *Catasticta* (currently under debate), *Ascia, Baltia, Belenois, Cepora, Charonias, Delias, Dixeia, Eucheira, Ganyra, Glennia, Glutophrissa, Hypsochila, Infraphulia, Itaballia, Leodonta, Leptophobia, Leuciacria, Melete, Mesapia, Mylothris, Neophasia, Pereute, Perrhybris, Phrissura, Phulia* (including *Piercolias, Pierphulia, Tatochila,* and *Theochila*), *Pieriballia, Pieris, Pontia* (including *Reliquia*), *Prioneris, Pseudanaphaeis, Saletara,* and *Sinopieris*

DISTRIBUTION
Worldwide except Antarctica and Greenland

HABITATS
The wealth of species is reflected in a high diversity of habitats, including wet and dry tropical forest, secondary forest, and forest margins from sea level up to some of the highest elevations in the Americas (13,000 ft/4,000 m)

as well as a combination of bifid pretarsal claws and hind leg tibiae with spurs. Males have specialized scales on the wings called androconia, with fine projections at the border of the wings (fimbriae, see p. 48). Wing coloration is predominantly white with black scales and patterns; many species also have yellow or orange scales on their wings. Species in the Neotropical genera *Archonias*, *Pereute*, and *Eucheira* are mostly

(fimbriae, see p. 48)

ABOVE | A female Painted Jezebel *Delias hyparete*, a common and widely distributed species. It has been reported that the caterpillars throw themselves to the end of a silk thread when they feel threatened by predators.

SIZE
Medium: ¾–1½ in (20–38 mm)

HOST PLANT FAMILIES
Plants in families Alliaceae, Anacardiaceae, Annonaceae, Arecaceae, Asteraceae, Balsaminaceae, Bataceae, Boraginaceae, Brassicaceae, Celastraceae, Crassulaceae, Dioscoreaceae, Ebenaceae, Euphorbiaceae, Fabaceae, Loranthaceae, Malvaceae, Phyllanthaceae, Pinaceae, Polygonaceae, Rhizophoraceae, Rubiaceae, Rutaceae, and Verbenaceae

CONSERVATION
The endemic Madeiran Large White *Pieris wollastoni* is categorized as Critically Endangered on the IUCN Red List. Despite recent fieldwork in remaining laurel forests, this butterfly has not been found since it was last recorded in 1986. The European Red List of Butterflies records this butterfly as the first European species to become globally extinct in recent times. The Black-veined White *Aporia crataegi* is now extinct in the UK, although it is still found in continental Europe. Other island endemics include the Canary Islands Large White *Pieris cheiranthi* and Madeira Brimstone *Gonepteryx maderensis*, both categorized as Endangered. The Wood White *Leptidea sinapis* is categorized as Endangered in the UK and the African species *Mylothris sagala* as Critically Endangered

black in the wings and could be confused with swallowtail butterflies. Males can be similar in appearance and size to females; however, there are remarkable examples of strong sexual dimorphism in Andean *Perrhybris pamela* and many Asian and Australian *Delias* including *D. chrysomelaena*. Some species are of great interest for scientists studying sexual signaling in butterflies; for example, members of the Palearctic genus *Pieris* have hidden ultraviolet patterns, with males having more ultraviolet-absorbent pigments in the wing scales than females, which are usually diffusely ultraviolet-reflectant.

Immature stages vary across the subfamily. Eggs can be laid singly or in clusters and larvae are elongated and cylindrical, some with hairs and large heads colored with black or red. Pupae can mimic excrement, such as those of genera *Perrhybris* and *Melete*. Caterpillars of the Cabbage White (known as Small White in Europe) *Pieris*

rapae spread by accident from Europe, causing significant damage to cabbage, kale, and broccoli crops in North America, New Zealand, and Australia. Mexican *Eucheira socialis* forms silk nests, and its caterpillars are farmed for consumption. Their striking combination of colors and patterns have made this group of butterflies very popular and precious among amateur collectors.

WHITES

Despite the vast diversity in the Pieridae family and the subfamily Pierinae, many groups remain poorly studied as they includes species with limited distribution in unique habitats. Other tribes in Pierinae include Anthocharidini, which contains seven genera, 65 species, and around 190 subspecies, with a cosmopolitan distribution in the Palearctic (*Zegris*), USA (*Anthocharis, Euchloe*), Africa, Asia (*Hebomoia*), and the Neotropical region (*Eroessa, Hesperocharis, Mathania*). Many species display intricate and spectacular patterns on the ventral side of the hindwings—for example,

RIGHT | The Orange-tip *Anthocharis cardamines* is a charismatic species widespread across Europe and temperate Asia. The distinctive orange bands on the forewings, which give the butterfly its name, are absent in females.

GENERA
Leptosia (Leptosiaini); *Nepheronia* and *Pareronia* (Nepheroniini); *Anthocharis*, *Eroessa*, *Euchloe*, *Hebomoia*, *Hesperocharis*, *Mathania*, and *Zegris* (Anthocharidini)

DISTRIBUTION
Calopierini is found in Sudan, western Saudi Arabia, and Yemen; Elodinini in Indo-Australia; Leptosiaini in Africa, Indonesia, and Australia; Nepheroniini in Africa and Indonesia; Teracolini in tropical Africa and Asia; and Anthocharidini is worldwide excluding the northern Palearctic, New Guinea, and New Zealand

HABITATS
The wealth of species is reflected in a high diversity of habitats, which include ridges and hilltops, sub-Saharan habitats, frost-free savanna, and forest margins from sea level up to 6,500 ft (2,000 m)

SIZE
Small to large: ¾–1½ in (20–38 mm)

the Neotropical species *Hesperocharis graphites* and *Eroessa chiliensis*, European Orange Tip *Anthocharis cardamines*, and North American *Euchloe lotta*, to name a few. *Hesperocharis* includes uncommon inhabitants of forest at 2,300–6,500 ft (700–2,000 m) elevation, flying along road cuttings. Caterpillars resemble bird droppings. Other uncommon species include *Euchloe bazae*, an endemic of Spain listed as Vulnerable to extinction.

The tribe Teracolini includes five genera, 67 species, and double the number of subspecies described to date. Species in *Eronia*, *Gideona*, and patterned *Pinacopteryx* are found only in Africa, while *Colotis* can also be found in the southwest of the Arabian Peninsula. The genus *Ixias* is only found in Asia. The African species *Pinacopteryx eriphia* has different forms in the wet and dry seasons, which look quite distinct from one another. Males and females of some species are similar in appearance and size.

Previous phylogenies and recent molecular revisions using morphological and immature characters have resulted in three individual genera having their own tribe: Calopierini (African genus *Calopieris*), Elodinini (Australasian *Elodina*), Leptosiaini (cosmopolitan *Leptosia*, found in Africa, Australia, and Indonesia). Nepheroniini includes two genera: *Nepheronia*, found in Africa, and Indo-Australian *Zebra*.

HOST PLANT FAMILIES
Brassicaceae (*Anthocharis, Euchloe, Zegris*), Capparaceae (Calopierini), Capparaceae (Elodinini, *Hebomoia, Ixias, Leptosia, Pareronia, Pinacopteryx*), and Salvadoraceae (*Colotis, Eronia, Nepheronia*)

CONSERVATION
There are no species outside Pierini categorized as threatened in the IUCN Red List

TOP | Caterpillars of this species *Colotis phisadia* feed on the "toothbrush tree" (*Salvadora persica*), known for its use in oral hygiene. The common name of this butterfly, Blue Spotted Arab, is confusing because it is the same for the Indian species *Colotis protractus*.

ABOVE | Species in the genus *Pareronia*, commonly known as Wanderers, are found in Southeast Asia and India. They are distinctive among other Oriental Pieridae because of the venation, with all standard veins present in their forewings.

RIODINIDAE
METALMARKS

Butterflies in this family are commonly called metalmarks because of the metallic and brightly colored scales that many of them have on their wings. Organized in over 160 genera, there are over 1,600 species and more than a thousand subspecies described. Until recently little was known about this family, which was formerly listed as Nemeobiidae or even included as a subfamily in the Lycaenidae. Researchers have calculated that the two families diverged around 88 mya, in the late Cretaceous period. The current number of genera, as well as tribe arrangements, varies greatly in published literature that includes comprehensive morphological and molecular phylogenies. Here we update and reconcile the most recent changes resulting from published research, conducted by different teams, into the taxonomy of the group.

The Riodinidae is arranged in two subfamilies, Nemeobiinae and Riodininae. Around 166 genera are included and are found mostly in the Neotropical region, with a few genera found in tropical Africa and Asia, and only one species in Europe. Butterflies in this family are very small and fly fast and high in the canopy of tall trees in forests during sunny hours. They are found from sea level to high elevations in mountains up to 8,200 ft (2,500 m); some species occur in secondary habitats. The forelegs are greatly reduced and non-functional in males but developed in females. Many species are sexually dimorphic and some display spectacular patterns in the wings.

Life histories and host plants are unknown or poorly studied in most species in this family, with large gaps in our knowledge of their ecology, distribution, and threat status.

Eggs can vary in shape and sculptural features. The caterpillars in Riodinidae have innumerable feeding behaviors from the usual consumption of leaves, nectar, or detritus to predation of smaller insects. The genus *Theope* has stridulatory organs, enabling them to produce calls and interact with ants in a symbiotic relationship known as myrmecophily. The caterpillars can be gregarious and are

BELOW | When fluttering around, the transparent wings of the Fabricius Angel *Chorinea octauius* reflect a dazzling blue resembling a large wasp. This species can be locally common in forested areas in tropical countries.

known to feed on a variety of plants from over 40 different families. Pupae vary across groups and species.

Specimens of extant species are scarce in scientific collections, because they are not only hard to catch but also small and fragile, and their wings do not spread easily when prepared for preservation. As a result, many groups in the Riodinidae family are understudied, with many new species still to be described and the immature stages of most species yet to be discovered. As most species of Riodinidae are not native to English-speaking countries, there are few common English names used for this family. Some Riodinidae fossils have been found of species in the genera *Lithopsyche*, *Lycaenites*, and *Riodinella*.

BELOW |
The metalmark *Ourocnemis renaldus* looks completely different from the spectacular patterns displayed on the underside of the wings. Its upperside is bright metallic blue with tiny white spots. This species is found from Nicaragua to Brazil.

HARLEQUINS, JUDYS, PUNCHES, AND METALMARKS

This subfamily includes most non-Neotropical species in the Riodinidae, arranged in two tribes, Nemeobiini and Euselasiini, 26 genera, and almost 300 species spread across the world. The majority of the 110-plus species in the tribe Nemeobiini occur in Southeast Asia, with only one genus found in the Australasian region (*Praetaxila*), two in Africa (*Afriodinia*, *Saribia*), and a single species in Europe, the Duke of Burgundy *Hamearis lucina*. Peruvian endemic *Styx infernalis*—classified previously as a moth or in other families because of its unusual appearance—and the Costa Rican Metalmark *Corrachia leucoplaga* are the only two Neotropical species in this tribe and are closer to relatives in the Old World than they are to their Neotropical relatives. Little is known about these unusual species, including their immature stages. Butterflies in Nemeobiinae are associated with dense, pristine forest. There are some records of crepuscular species, such as Australia's *Praetaxila segecia*. Males and females are alike, with only small differences in appearance. Madagascan endemics *Saribia decaryi*, *S. ochracea*, *S. perroti*, and *S. tepahi* are not protected or listed, and should be

BELOW | *Hamearis lucina*, known as the Duke of Burgundy, is the only species in the Riodinidae family found in Europe.

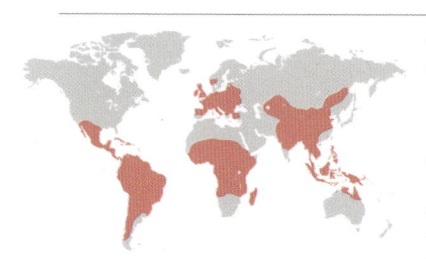

GENERA
Abisara, Afriodinia, Dicallaneura, Laxita, Paralaxita, Praetaxila, Saribia, Stiboges, Taxila, Corrachia, Dodona, Hamearis, Polycaena, Styx, Takashia, Tibetododona, and *Zemeros* (Nemeobiini); *Erythia, Eugelasia, Eurylasia, Euselasia, Hades, Maculasia, Marmessus, Methone, Myselasia,* and *Pelolasia* (Euselasiini)

DISTRIBUTION
Nearctic and Palearctic regions, Asia, and Neotropics, with two genera occurring in Africa and one each in Europe and Australasia

HABITATS
Tropical montane cloud and rainforests, and African lowland forested hills up to 4,000 ft (1,200 m); Neotropical forests at 3,300–5,200 ft (1,000–1,600 m) elevation; Palearctic chalk and limestone grassland with scrub, and clearings in ancient woodland

assessed because of current accelerated deforestation in Madagascar.

The tribe Euselasiini comprises over 187 species that have a distinctive junction in the forewing's veins. They are distributed only in the Neotropics. This tribe is regarded by some authors as a subfamily. It formerly contained three genera until new ones were described more recently, using primarily molecular characters. However, species in the Neotropical *Euselasia* and related genera need further comprehensive revision. Most species in Euselasiini show marked sexual dimorphism; females look similar to each other and they are often confused with species in other families too.

SIZE
Small to medium: ⅔–1 in (15–25 mm)

HOST PLANT FAMILIES
Myrsinaceae (*Praetaxila, Abisara*), Primulaceae (*Styx, Corrachia*), Clusiaceae, and Melastomataceae (*Euselasia*)

CONSERVATION
Only a few species in Nemeobiinae have been assessed and these are categorized in the IUCN Red List as Least Concern. However, the Duke of Burgundy *Hamearis lucina* has shown a sharp 84% decline in its population in the UK since the 1970s

ABOVE | Little is known about the biology of the beautiful *Saribia tepahi*. It is endemic to Madagascar and inhabits forested areas.

OPPOSITE BELOW | Until recently, the metalmark *Erythia thucydides* was placed in the large genus *Euselasia*. This uncommon small butterfly is only found in Brazil.

METALMARKS

This subfamily is the most species-rich in Riodinidae, with all species found in the Neotropical region. These small, charismatic butterflies have attracted considerable attention for recent study. Riodininae includes ten tribes, around 136 genera, and more than 1,316 species, with many more yet to be described. Numbers and taxonomic arrangements vary according to the morphological, ecological, and molecular characters used by different research teams. The immature stages of most species are unknown.

The newly described tribe Befrostiini includes only one genus, the Amazonian *Befrostia*. The tribes Callistiumini and Dianesiini also have

LEFT | *Sarota* includes some of the smallest, most charismatic butterflies from Mexico through Central and South America. The intricate metallic patterns on the underside of the wings contrast with the plain dark colors on the upperside.

GENERA
Befrostia (Befrostiini); *Callistium* (Callistiumiini); *Calydna* (Calydnini); *Dianesia* (Dianesiini); *Alesa, Eurybia, Cremna, Eunogyra, Hyphilaria, Ithomiola* (and *Hermathena*), *Leucochimona, Mesophthalma, Mesosemia, Napaea, Perophthalma, Semomesia, Teratophthalma, Voltinia, Eucorna* (Eurybiini); *Helicopis* (Helicopini); *Xanthusa* (Sertoniini); *Argyrogrammana, Asymma, Chimastrum, Esthemopsis, Lucillella, Mesene, Mesenopsis,* *Mycastor, Phaenochitonia, Pirascca, Pterographium, Stichelia, Symmachia, Tigria, Xenandra,* and *Xynias* (Symmachiini)

DISTRIBUTION
Neotropical, Caribbean, and southwestern North America; some members of the Emesidini range into the temperate zones of the USA and close to or into Canada

HABITATS
Wet montane forests, rainforest, forest edges, and streams from sea level up to

only one species each, Brazilian endemic *Callistium cleadas* and Caribbean Metalmark *Dianesia carteri*, respectively.

The tribe Sertaniini contains butterflies restricted to xeric habitats, including the recently described genera *Sertania* and *Xanthosa*. The tribe Calydnini includes agile butterflies with patterned wings and dark ground colors; some in the genera *Calydna*, *Echenais*, *Echydna*, and *Imelda* are covered in small spots.

The tribe Emesidini includes widespread species that feed from several families of plants, including the genera *Apodemia*, *Curvie*, and *Emesis*. The first of these ranges into North America. The tribe

Helicopini contains small, appealing butterflies in the genera *Anteros*, *Helicopis*, *Ourocnemis*, and *Sarota* that show bright colors and the metallic wing markings Riodininae are known for.

Species in the tribe Mesosemiini that lack a spur on the tibia and have hairy eyes are now included in the tribe Eurybiini. The tribe Symmachiini includes small, brightly colored butterflies, particularly those in the genera *Lucillella* and *Xenandra*, and butterflies with unusual wing shapes or metallic scales, as in *Symmachia* and *Argyrogrammana*, respectively. Males of species in Symmachiini have androconial scales. The caterpillars are covered in short bristles.

8,200 ft (2,500 m); a few species occur in open areas

SIZE
Small to medium: ⅓–1⅓ in (10–35 mm)

HOST PLANT FAMILIES
Euphorbiaceae, Melastomataceae, Ulmaceae, Marantaceae, Zingiberaceae, and Solanaceae. Host plants are unknown for most species

CONSERVATION
No species in this subfamily have been categorized by the IUCN. A few species (e.g., Costa Rican *Sarota turrialbensis* and *Voltinia theata*) are known only from collections, suggesting they may be rare, and studies are needed to determine whether they are still found in the wild

METALMARKS

This tribe of small Neotropical butterflies contains over 45 genera, and almost 400 species and 95 subspecies. Adults display a wide range of patterns, sizes, and wing shapes. Among the more unusual members of Nymphidiini are *Rodinia calphurnia*, which displays a tail-like hindwing projection and *Zelotaea grosnyi* and *Behemothia godmanii*, which have unusually pointy forewings. The genus *Theope* contains brightly colored species, while *Stalachtis* contains species that are bright orange with tiger-like patterns. Others, like Brazilian endemic *Zabuella paucipuncta*, are dull. Adults are found in moist to wet habitats: *Periplacis* species, for example, have been observed flying over beaches. Sexual dimorphism varies across genera: males and females of *Juditha* and *Nymphidium* are alike, but those of *Synargis* and *Calospila* are distinct. Males have been observed perching in groups at the top of trees. Caterpillars of some species in *Setabis* and *Calospila* are tended by ants.

TOP | The metalmark *Synargis phylleus* is found from Costa Rica to Colombia. Species in this group of small butterflies display strong dimorphism, with females sometimes described as different species.

ABOVE | This uncommon species *Thisbe ucubis* inhabits rainforests and mimics toxic day-flying moths in the genus *Hypocritta*. Its caterpillars hide during the day and are active at night, when they feed.

GENERA INCLUDE
Minstrellus, Pachythone, Zabuella, Adelotypa, Archaeonympha, Argyraspila, Ariconias, Aricoris, Calociasma, Catagrammina, Catocyclotis, Hypophylla, Joiceya, Lemonias, Livendula, Minotauros, Pandemos, Protonymphidia, Pseudolivendula, Pseudonymphidia, Pseudotinea, Sanguinea, Setabis, Thenpea, Zelotaea, and Machaya (including Pachitone)

DISTRIBUTION
Neotropical, a few species into North America

HABITATS
Pristine forest, forest edges and streamsides; ridge tops from low elevations up to 4,500 ft (1,400 m)

SIZE
Small to medium: ⅓–1⅓ in (8–33 mm)

HOST PLANT FAMILIES
Melastomataceae, Euphorbiaceae, Fabaceae, Malvaceae, Marcgraviaceae, Malpighiaceae, Dilleniaceae, Clusiaceae, Urticaceae, Convolvulaceae, Lecythidaceae, and Sapindaceae

CONSERVATION
Brazilian endemic Joiceya praeclarus is categorized as Endangered on the IUCN Red List

METALMARKS AND GLASSWINGS

This large tribe in the Riodinidae includes 42 genera, almost 300 species, and 270 subspecies. Butterflies in this tribe are fast, flying high in the canopy. Most occur in the Neotropics, with *Calephelis* present in North America. Wing shapes and colors vary, with some species displaying tails, such as *Syrmatia aethiops* and *Barbicornis basilis*. Males congregate in groups, hilltopping and perching in bushes. Seasonal species in some genera, for instance in *Calephelis*, have adults that vary in color, size, or shape according to the different seasonal conditions into which they will emerge, making identification complex. Sexes present slight dimorphism, although in colorful genera, such as *Ancyluris* and *Lasaia*, males are iridescent blue while females are not.

Eggs are laid singly. The caterpillars of the majority of known species, including Central American *Brachyglenis dinora* and *Rhetus jurgensenii*, have very long cetae and are easily confused with caterpillars of moths. Caterpillars of some species are tended by ants. There are records of caterpillars of species in *Melanis* being attacked by tachinid flies and pupae being attacked by parasitic chalcid wasps. However, little is known about immature stages in most species.

ABOVE | Most species in the spectacular Neotropical genus *Caria* are recognized by the vivid metallic green on the upperside of the wings (in males) and the markedly bent costa in the forewings. This is *Caria castalia*.

GENERA INCLUDE
Amarynthis, Baeotis, Calephelis, Caria, Cariomothis, Cartea, Chalodeta, Chamaelimnas, Charis, Chorinea, Crocozona, Dachetola, Detritivora, Exoplisia, Isapis, Ithomeis, Lyropteryx (including *Necyria*), *Metacharis, Panara, Pheles, Riodina, Siseme,* and *Themone*

DISTRIBUTION
Neotropical; a few into temperate North America

HABITATS
Predominantly in lowlands, in forest and forest edges, and streamsides; ridge tops, some in mountains up to 4,500 ft (1,400 m)

SIZE
Small to medium: ⅓–1 in (9–28 mm)

HOST PLANT FAMILIES
Melastomataceae, Euphorbiaceae, Fabaceae, Malvaceae, Sapindaceae, Passifloraceae, Bromeliaceae, Ulmaceae, Liliaceae, Malpighiaceae, Poaceae, and Celastraceae

CONSERVATION
Brazilian endemic *Rhetus belphegor* (in genus *Nirodia* until recently) is categorized as Endangered on the IUCN Red List

LYCAENIDAE
HARVESTERS, BLUES, HAIRSTREAKS, AND COPPERS

Lycaenidae is the second-largest family of butterflies after Nymphalidae, comprising over 5,200 species and a similar number of subspecies, arranged in approximately 470 genera, 42 tribes, and six subfamilies. Species in this family are found in every continent except Antarctica in a wide variety of habitats from pristine tropical forests to African savanna, and isolated mountains to secondary areas. In some parts of the world, including Africa, Lycaenidae is the most diverse group of butterflies. The vast species diversity means the taxonomy of this group is yet to be resolved. Recent studies combining morphological, ecological, and molecular characters have advanced knowledge of this poorly studied family. However, the highest diversity occurs in the Neotropics, while most of the research has involved African and Asian species.

Identifying species can be very complex as many are alike, and marked sexual dimorphism in many species makes it difficult to associate the sexes. The presence of specialized and modified scales in the forewings of males is a valuable diagnostic character for identification of species and relationships among species. Their small size and fast flight in tree and shrub tops make these elusive butterflies hard to collect.

This group is most noteworthy for its peculiar ecology, with the immature stages being associated with ants. This relationship may have begun as symbiotic defense, but it makes this group susceptible to changes in the environment. The Curetinae has a loose association with ants. The adults often feed on torn leaves, possibly torn by ants, and the immature stages emit secretions that may placate the ants. Species in other subfamilies—such as the African Aphnaeinae, the globally distributed Lycaeninae, and most Theclinae—have a mutualistic relationship with other species. The soft, slug-like caterpillars are easy prey for predators. However, they have glands that secrete "honey"—acquired from host plant sap—which attracts ants, which, in turn, protect the caterpillars. Caterpillars and ants associate closely while feeding, and in some cases the caterpillars are protected by the ants, which bring them into the nest and may even allow pupation inside the ant nest. The subfamily Miletinae and the Theclinae tribe Polyommatini have benefited from parasitic relationships, with caterpillars often feeding on the ant brood. In contrast, members of the African subfamily Poritiinae display an unusual ecology, feeding principally on lichens and sometimes even laying their eggs not on plants, as most butterflies, but on surfaces such as rocks and walls in houses.

The Lycaenidae includes small, blue butterflies as well as some with brown, white, orange, or yellow colors in the wings. Adult antennae are placed close to the eyes, often resulting in an indentation. With such a diverse range of ecologies, there are few obvious unifying characters for immature stages and eggs. However, many caterpillars lack hairs or other physiological features that deter predators, relying instead on their association with ants for protection. Eggs are usually domed, often have a depression in the center, and are densely ridged.

OPPOSITE | The African Checkered Gem *Zeritis neriene* flies closer to the ground in savanna and cleared grounds.

OPPOSITE | The spectacular male of *Simiskina phalia*, an uncommon species found in semi-dry forests in lowlands in Southeast Asia. This species displays marked sexual dimorphism, with females looking dull orange, strongly contrasting with the metallic blue colored males.

SUNBEAMS

This small group of lycaenids is the least understood and is represented only by the genus *Curetis* with 18 species and approximately 44 subspecies known. Unlike many other Lycaenidae, they do not have a close symbiotic relationship with ants but ants have adapted to tolerate them, perhaps linked to them sharing similar food.

Adults have an unusual sensory organ made of bristles on the antennae shafts that is not found in other Lycaenidae. Wings of butterflies in this group are white or grayish on the underside and usually orange with black borders on the upperside (except for one species, whose female is white with black borders). They lack tailed hindwings and show differences in wing venation from the other Lycaenidae subfamilies. Adults feed in damp areas and especially on damaged leaves. Caterpillars of *Curetis* species are phytophagous and do not have the honey gland of other Lycaenidae groups. Instead, the caterpillars have sclerotized cylinders or modified tentacle-like structures that are everted when in danger. This is a unique feature not found in another Lycaenidae.

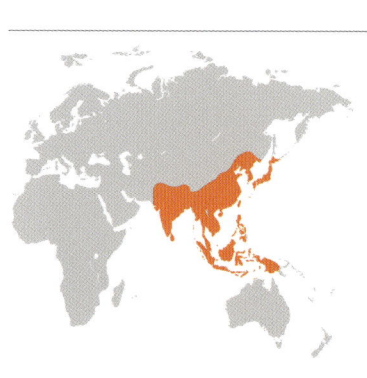

Eggs are flattened spheres with hexagonal lines. Females lay their eggs on the underside of leaves at the base of bushes. Caterpillars are small and slug-like, with perforated cupola organs clustered around the spiracles, which are thought to be related to their interactions with ants. There are published observations of caterpillars of the Asian species *Curetis regula* being constantly attended by ants when walking to their feeding sites and of adults feeding on damaged tissue of host plants.

GENERA
Curetis

DISTRIBUTION
South and Southeast Asia

HABITATS
Rainforests

SIZE
Small: ⅓–½ in (8–11 mm)

HOST PLANT FAMILIES
Fabaceae and Mimosaceae

CONSERVATION
No species in this subfamily have been assessed using IUCN Red List criteria

OPPOSITE | Many species within the Asian genus *Curetis*, known as sunbeams, have a distinctive plain silvery white underside. However, the very similar patterns make species difficult to distinguish based solely on appearance, and study of genitalia helps accurate identification.

LEFT | A female *Curetis bulis*. This species is the only one in the Malay Peninsula whose females have this brown-and-white appearance contrasting with the bright orange wings in the male.

BELOW | Males in the genus *Curetis* are recognizable for their orange wings with black borders on the uppersides. They are often found in moist areas such as riverbanks and forest paths.

HARVESTERS AND WOOLLY LEGS

The Miletinae is arranged in five tribes and comprises 13 genera and almost 200 species, with a similar number of subspecies described. This subfamily is most diverse in Asia and Africa, with the genus *Liphyra* found in Australasia and the genus *Feniseca* in North America. Adults lack spurs on the mid and hind legs; in males the front legs are fully functional. There are five recognized tribes, including the previously recognized subfamily Liphyrininae, now a tribe (Liphyrini), all of which have different wing shapes and coloration as well as ecologies. They are mostly small and dull with browns and whites dominating, although some Liphyrini include brighter oranges. The Asian genus *Taraka* includes small butterflies with white ground color and black dots.

Butterflies in this subfamily have fascinating immature stage ecologies in that their caterpillars are almost exclusively insectivorous, although some drink from ant secretions. This behavior is shared only with some species in the subfamily Poritiinae. Caterpillars usually feed on adult aphids, mealybugs, and leafhoppers (all Hemiptera) as well as ants (Hymenoptera: Formicidae)—ants often harvest Hemiptera too. Non-ant-eating Miletinae often still have a close association with ants. Others cover themselves with the secretions of another species (such as the woolly secretions of mealybugs). Most use chemicals that prevent their prey species identifying them. Some caterpillars live inside ant nests and feed on the larvae. Pupae will then often develop within the ant nest. Caterpillars and pupae can have a leathery or tough outer skin, probably to protect them from cohabiting insects.

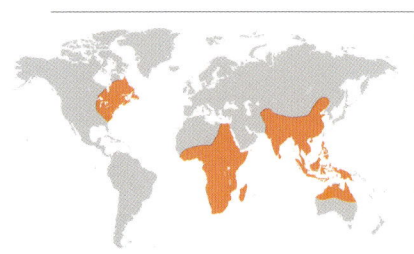

GENERA
Euliphyra, Aslauga, Liphyra (Liphyrini); *Lachnocnema, Thestor* (Lachnocnemini); *Taraka* (Tarakini); *Spalgis, Feniseca* (Spalgini); *Miletus, Megalopalpus, Allotinus, Logania,* and *Lontalius* (Miletini)

DISTRIBUTION
Mostly tropical Asia; also, Africa, Australasia, North America

HABITATS
Tropical rainforests, savanna, and coastal forests; temperate forests in North America

SIZE
Small to medium: ½–¾ in (15–21 mm)

HOST PLANTS
None; they feed on other insects

ABOVE | The Harvester *Feniseca tarquinius* is one of the most interesting and best studied butterfly species, also the only species with carnivorous caterpillars in North America.

OPPOSITE | The Common Woolly Legs *Lachnocnema bibulus* is a tiny butterfly found in savanna and open forest in sub-Saharan Africa. Its caterpillars often pupate in the superficial galleries of ant nests.

RIGHT | The charismatic Forest Pierrot *Taraka hamada* is a predator of greenflies during its caterpillar stage. Its interesting lifecycle has attracted attention to commercialize it as a pet in South Korea.

CONSERVATION
South African endemic *Aslauga australis* is categorized in the IUCN Red List as Endangered. Habitat deterioration is the main threat for this species because of house-building, logging, and intense overgrazing. The Indian subspecies Forest Pierrot *Taraka hamada imperialis* is legally protected in India under Schedule II of the Wildlife (Protection) Act

SILVERLINES

BELOW | Stunning common high flyer *Aphnaeus orcas* is found in almost all of Africa and accounts for the majority of specimens of this genus found in museum collections. However, various new species have been named during the last decade.

Aphnaeinae was for years treated as a tribe within Lycaeninae or Theclinae. However, following recent molecular studies it has been universally redefined as a distinct and valid separate subfamily with three tribes, Axiocersini, Aphnaeini, and Garitini. There are around 15

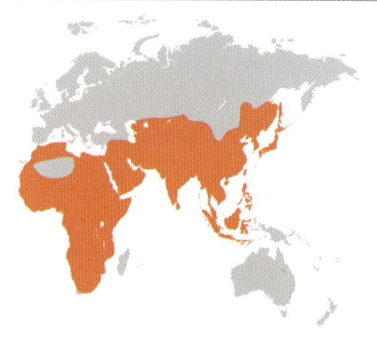

GENERA
Axiocerses, Zeritis (Axiocersini); *Aphnaeus, Tylopaedia, Phasis, Trimenia, Argyraspodes, Erikssonia, Aloeides* (Aphnaeini); *Pseudaletis, Chrysoritis, Vansomerenia, Chloroselas, Lipaphnaeus, Crudaria, Cesa,* and *Cigaritis* (Cigaritini)

DISTRIBUTION
Mainly African, but some species occur in Asia

HABITATS
Montane grassland, sandy patches, rocky ridges, shrubs, and near streams; at elevations up to 5,200 ft (1,600 m)

SIZE
Small to medium: ⅔–1 in (15–24 mm)

HOST PLANT FAMILIES
Anacardiaceae, Asteraceae, Caesalpiniaceae, Euphorbiaceae, Fabaceae, Melianthaceae, Mimosaceae, and Zygophyllaceae

LEFT | Common Silverline *Cigaritis vulcanus* is found in India and South and Southeast Asia. This species was found and named 250 years ago. It is sometimes listed under the genus *Spindasis*.

recognized genera, and approximately 298 species, with a similar number of subspecies; more remain to be described, mainly in hotspots and unexplored areas, such as the Indian subcontinent. Most species are inhabitants of sub-Saharan habitats on the African continent, but the genus *Cigaritis* is found in Asia and the southern Palearctic region.

Adults of most species have very thin, tail-like processes on the hindwing and a dorsal nectary organ that secretes honeydew, which is attractive to ants, with which many species have a close association. Most species have silvery markings on the underside of their pale wings, a feature shared only with some Theclinae, and which provides their common name. The upperside of the wing is usually brown or orange. Eggs are dome-shaped. Caterpillars are narrow with a small head.

The relationship of this family with ants is not parasitical but mutualistic. Caterpillars feed on plants and the ants defend the caterpillars from predators. The caterpillars exude secretions that contain nutrients, which the ants feed on. Pupae tend to be located in sheltered areas.

Populations of South African endemic species in the genera *Aloeides* and *Chrysoritis* are facing drastic host plant and habitat reduction due to the spread of invasive alien plants and inappropriate fire regimes.

CONSERVATION

South African species *Aloeides rossouwi*, *Aloeides stevensoni*, *Chrysoritis dicksoni*, and *Trimenia wallengrenii* are categorized as Critically Endangered and *Chrysoritis aureus*, *Chrysoritis adonis*, and *Chrysoritis rileyi* as Endangered on the IUCN Red List

GEMS AND WHITE MIMICS

The Poritiinae subfamily, found in Africa and Asia, has a large diversity of species and unique feeding habits. There are 58 genera, around 735 species, and 410 subspecies organized into five tribes. The caterpillars of most species feed not on plants but on algae, fungi, and lichens. Some feed on ant larvae, having similar ecology to the Miletinae.

Adults are small and their hindwings lack tails. Males can be brown, orange, green, or blue, often in bright shades; females are more variable but usually have dark wing borders. Some species in the genera *Falcuna* and *Ornipholidotos* are white with elaborate black patterns on the underside of the hindwings and show strong sexual dimorphism. Species in the genus *Alaena* are yellow with strong black markings in the wing veins.

Many species associate with trees used regularly by ants for nesting or feeding and adults probably live in their nests, likely for protection rather than to parasitize them. Some adults display an unusual behavior of extending their wings down to below the abdomen in the presence of ants, which may be to exude chemicals as a form of communication.

The tribes Liptenini and Pentilini were previously placed in a separate subfamily, Lipteninae. The immature stages eat algae, fungi, and lichen, usually on trees. Many species in the

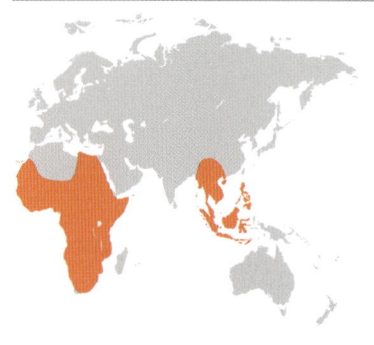

GENERA
Genera include: *Cyaniriodes, Deramas, Simiskina, Poritia,* and *Poriskina* (Poritiini); *Cooksonia, Durbaniopsis,* and *Durbaniella* (Mimacraeini); *Telipna, Ornipholidotos, Torbenia,* and *Pentila* (Pentilini); *Phytala, Hewitsonia, Hypophytala, Neoepitola, Cephetola, Geritola, Stempfferia, Powellana, Iridana, Batelusia, Hewitola, Pseudoneaveia, Cerautola, Epitola, Aethiopana, Neaveia, Deloneura,* and *Tumerepedes* (Epilotini); *Tetrarhanis,*

Micropentila, Epitolina, Liptena, Teriomima, Euthecta, Baliochila, Cnodontes, Eresinopsides, Congdonia, Parasiomera, Eresiomera, Argyrocheila, Eresina, Toxochitona, Pseuderesia, Mimacraea, and *Mimeresia* (Liptenini)

DISTRIBUTION
Asia and Africa

tribe Mimacraeini are associated with rock-growing lichens and lay their eggs on stones and bricks.

Caterpillars have long bristles, lack honey glands, and do not exude secretions (either as food or camouflage) to ants. Pupae lack girdles. The immature stages of many species remain unknown.

OPPOSITE LEFT | Butterflies in the African genus *Ornipholidotos* are commonly known as glasswings or white mimics. It is a complex group with more than 60 recognized species, most of them described over the last 20 years.

OPPOSITE RIGHT | A male of the Yellow Zulu *Alaena amazoula*, a species found in rocky areas in grassy savanna across Afrotropical countries. Its caterpillars feed on lichens.

HABITATS
Primary forests, pristine grasslands, forest undergrowth, rocky slopes, and riverine forest

SIZE
Small to medium: ⅓–¾ in (10–20 mm)

HOST PLANT FAMILIES
Poritiini: Combretaceae, Dipterocarpaceae, and Fagaceae. Liptenini: other organisms such as lichens, algae, and fungi; several feed on ant larvae

CONSERVATION
South African endemic *Deloneura immaculata* is now categorized as Extinct following various unsuccessful attempts to find live specimens. Also South African, *Alaena margaritacea* is categorized as Critically Endangered on the IUCN Red List. This species is known from only two severely fragmented localities that are not connected and outside protected areas

TOP | A male of the Common Gem *Poritia hewitsoni* basking on leaves. This species is found in Southeast Asia and in India, where it is legally protected.

ABOVE LEFT | A male *Telipna semirufa* known as Western Telipna, a species found in rainforests in West Africa.

ABOVE RIGHT | Common Dots *Micropentila adelgitha* gets its common name from the distinctive patterns on its wings.

COPPERS AND SAPPHIRES

This subfamily, which, as redefined in recent molecular studies, now includes only five genera, has over 117 species and 232 subspecies known—though there are probably more to describe because their habitats are isolated and complex. Commonly known as coppers, butterflies in the Lycaeninae are small and generally orange and brown, with some displaying bluish-violet shades on the upperside of the wings; the underside can be white with dotted or very colorful, intricate patterns of yellow or red. Some species in the Southeast Asian genera *Heliophorus* and *Melanolycaena* have delicate tailed hindwings. Species in the genus *Lycaena* are distributed across the globe, with most occurring in the Holarctic and a few in the Afrotropical region. The Guatemalan Copper *Iophanus pyrrhias* is the only species in the Neotropics.

Caterpillars of this subfamily have a close symbiotic relationship with ants, giving off secretions

LEFT | Butterflies in the *Heliophorus epicles* complex are known as Purple Sapphires. This species is found in Nepal and Southeast Asia. Its bright red caterpillars feed on buckwheat plants (*Rumex*).

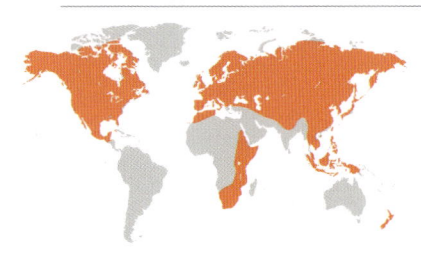

GENERA
Athamanthia, Heliophorus, Iophanus, Lycaena, and *Melanolycaena*

DISTRIBUTION
All continents except Antarctica and South America; primarily in the Holarctic region and Asia

HABITATS
Damp areas, fens, edges of ditches, upstream sections of rivers, stream banks, montane forests, semi-deserts, arid stony

foothills with xerophytic vegetation, marshy areas, and grassland; from sea level up to 10,000 ft (3,000 m)

SIZE
Small to medium: ½–¾ in (12–20 mm)

HOST PLANT FAMILIES
Mostly Polygonaceae; also Rhamnaceae, Ericaceae, Rosaceae, Fabaceae, and Plumbaginaceae

as they feed on plants, which the ants in turn feed upon. In return, the ants protect the caterpillars and often carry them back to their nests overnight to avoid predators. Some species pupate in ant nests. Pupae are usually girdled. The caterpillars are small and often green, with only small hairs and specialized glands (a dorsal nectary organ) that exude the secretions that ants consume—like some Aphnaeinae, although there are differences. Researchers have revealed that the host plant of the African species *Lycaena clarki* frequently grows in water, and caterpillars have adapted to survive submersion when feeding from it.

TOP | The rare Guatemalan Copper *Iophanus pyrrhias* is found in particular habitats at high elevations on volcanoes.

ABOVE | A female Large Copper *Lycaena dispar*. This species is extinct in the UK, and European populations are endangered because of the rapid conversion of its native wetland habitats to agriculture.

BLUES, HAIRSTREAKS, CHECKERED GEMS, SCARLETS, AND LEOPARD SILVERLINES

RIGHT | The striking Common Tinsel *Catapaecilma major* displays metallic scales on the underside of its wings. Its caterpillars live in ant nests and are escorted by ants during both feeding and movement on their host plant.

Recent higher classification studies have resulted in the Theclinae now comprising over 373 genera, more than 3,800 species, and around 4,000 subspecies arranged in 29 tribes. Tribes have different host plants within 50 families, including orchids, stonecrops, and ferns. Up to three tiny tails are usually present on the hindwing, varying in length between the tribes. Male butterflies often have androconial organs on one or both wings, a feature not found in the otherwise similar subfamily Aphnaeinae.

Members of tribe Oxylini have "false heads," the tails appearing like the body of the butterfly to confuse predators. Species in the tribes Pseudalmenini, Jalmenini, Candalidini, and Ogyrini are endemic to the Australasian region and include populations threatened by the degradation of mangroves and heathlands. Tribe Drinini includes only the Southeast Asian endemic genus *Drina*. Tribe Theclini includes *Hypaurotis crysalus*, the only Nearctic species not in the diverse Eumaeini, as well as *Laeosopis roboris*, one of the few species occurring in Europe. Males in the African genus *Iolaus* (Iolaini) appear to display hilltopping behavior, as do those of the Australian genus *Acrodipsas* (Luciini).

Immature stages of most species have associations with ants. Commonly these are mutualistic with species trading defense by ants for

GENERA
Genera include: *Pseudalmenus, Jalmenus, Candalides, Philiris, Hypochrysops, Titea, Ogyris, Drina, Tomares, Zesius, Iraota, Myrina, Amblypodia, Surendra, Lucia, Rapala, Deudorix, Pilodeudorix, Capys, Catapaecilma, Loxura, Horaga, Cheritrella, Drupadia, Tajuria, Iolaus, Dacalana, Oxylides, Remelana, Hypolycaena, Anthene, Cupidestes, Shirozua, Thecla, Chrysozephyrus,* and *Favonius*

DISTRIBUTION
Mostly tropical Asia and Africa. All continents except Antarctica and South America; also absent from New Zealand

HABITATS
All major biomes including savanna, mangroves, heathland, grassland, rainforest, and woodlands

SIZE
Small to medium: ½–¾ in (12–20 mm)

the food secretions of caterpillars—for example, caterpillars of *Hypochrysops* and *Surendra* are attended by green-tree ants. Some caterpillars, like those of *Arhopala*, are myrmecophagous, feeding on immature stages of ants. Eggs have ridges on the surface. Species in tribe Deudorigini lay eggs singly. Although immature stages are well-studied in some regions, most tropical species are understudied.

HOST PLANT FAMILIES
Betulaceae, Caesalpiniaceae, Crassulaceae, Dipterocarpaceae, Ericaceae, Euphorbiaceae, Fabaceae, Fagaceae, Juglandaceae, Lauraceae, Loranthaceae, Mimosaceae, Moraceae, Myrtaceae, Oleaceae, Orchidaceae, Oxalidaceae, Rosaceae, Rubiaceae, Solanaceae, and Verbenaceae

CONSERVATION
South African endemic *Capys penningtoni* is categorized as Critically Endangered, while Australian endemic *Acrodipsas illidgei*, Bathurst Copper *Paralucia spinifera*, and *Ogyris halmaturia* are categorized as Endangered on the IUCN Red List

TOP | A male of the Peacock Royal *Tajuria cippus*. This species is found from India across South and Southeast Asia.

ABOVE | A male of the Branded Imperial *Eooxylides tharis* displaying its "false-head" tails. This species is found in Southeast Asia.

HAIRSTREAKS, SCRUB-HAIRSTREAKS, AND GRAY HAIRSTREAKS

The Eumaeini is one of the most species-rich tribes. It is arranged over 104 genera and contains around 5 percent of the world's butterfly diversity with over 1,071 species and around 157 subspecies described. Over 200 species are yet to be described and have their higher classification settled.

Their morphological similarities make butterflies in this group almost impossible to tell apart without detailed examination, challenging taxonomists for almost 300 years. Molecular studies have estimated an age of approximately 30 million years for this tribe, one of the youngest recognized among butterflies. The hairstreaks include various attractive species displaying bright metallic blue in the dorsal wings of the males and an array of different patterns and colors in the undersides, such as the spectacular South American species *Evenus*

LEFT | The spectacular *Evenus felix* remained "hidden" within its closest relative, *Evenus coronata*, for over a century before it was finally named early this century.

OPPOSITE TOP | The Gray Hairstreak *Strymon melinus* is native to the Americas where it is widespread. However, it is believed to have been accidentally introduced to the UK, where it has also been recorded.

OPPOSITE MIDDLE | The Green Hairstreak *Callophrys rubi* is a widespread butterfly named for the metallic green scales displayed on the underside of its wings.

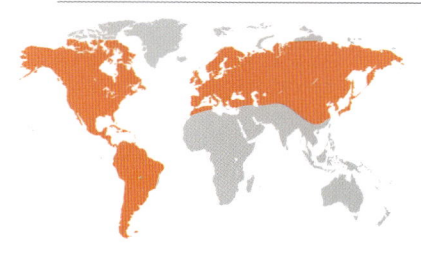

GENERA
The most species-diverse genera include: *Arcas, Arzecla, Badecla, Brangas, Brevianta, Callophrys, Camissecla, Contrafacia, Crimsinota, Denivia, Electrostrymon, Erora, Gargina, Gigantorubra, Iaspis, Ignata, Janthecla, Johnsonita, Lamprospilus, Laothus, Lathecla, Mercedes, Micandra, Ministrymon, Nesiostrymon, Nicolaea, Ocaria, Oenomaus, Olynthus, Panthiades, Paraspiculatus, Parrhasius, Penaincisalia, Podanotum, Rhamma,* *Strephonota, Symbiopsis, Theclopsis, Thereus, Timaeta,* and *Tmolus*

DISTRIBUTION
Mostly Neotropical; some genera found in the Holarctic region (*Atlides, Callophrys, Erora, Satyrium,* and *Strymon*) and Southeast Asia (*Satyrium*)

HABITATS
Rainforests, cloud forests, deserts, lowlands; some species found in mountains at elevations up to 13,000 ft (4,000 m)

coronata. The majority of species are distributed in the Neotropical region, with about 10 percent found in the Holarctic region, including gray hairstreaks in the genus *Strymon*. In the large majority of species males possess at least one secondary sexual organ, and many have delicate tail-like features on the hindwings. Dimorphism is notable with females dull in coloration and with much rounder wings. It is estimated that lifecycles remain unknown for around 75 percent of species in this tribe. Eggs of some species of *Calycopis* are reported to be laid in fallen leaves or seeds, suggesting they are detritivores and feed on decaying organic matter, unlike most caterpillars, which feed on live plants (phytophagy). As with other Theclinae, caterpillars associate with ants in mutualistic relationships, likely to benefit from the protection received.

SIZE
Small to medium: 1/3–1 1/2 in (8–37 mm)

HOST PLANT FAMILIES
Zamiaceae, Sapotaceae, Cunoniaceae, Solanaceae, Bromeliaceae, Anacardiaceae, Combretaceae, Fabaceae, and Loranthaceae; some caterpillars (e.g., *Callophrys*) feed on seeds and mushrooms

CONSERVATION
Brazilian endemics *Arawacus aethesa*, *Strymon ohausi*, and *Magnastigma julia* are listed as Endangered in the Red Book of Endangered Fauna of Brazil

ABOVE LEFT | Gray Ministreak *Ministrymon azia* displaying the characteristic patterns on the underside of its wings. This is probably one of the most complex tribes of butterflies to identify.

ABOVE RIGHT | The Red-banded Hairstreak *Calycopis cecrops* is a very common species found in southeastern USA and Mexico.

BLUES, CUPIDS, PIERROTS, AND GINGER WHITES

ABOVE | Males of spectacular Adonis Blue *Polyommatus bellargus* contrast with the drab brown colored females. This species is found from Europe through the Palearctic region.

RIGHT | Species in the genus *Itylos* are small butterflies found in the southern latitudes of South America. They resemble Palearctic species in appearance.

GENERA
The most species-diverse genera include: *Polyommatus, Lepidochrysops, Jamides, Plebejus, Pseudolucia, Nacaduba, Udara, Agriades, Euchrysops, Itylos, Celastrina, Leptotes, Uranothauma, Tarucus, Cupido, Kretania, Prosotas, Danis, Tongeia, Lysandra, Eicochrysops, Harpendyreus Thermoniphas, Glaucopsyche, Euphilotes, Turanana, Chilades, Orachrysops, Azanus, Pseudophilotes,* and *Tuxentius*

DISTRIBUTION
Worldwide except Antractica

HABITATS
Tropical rainforests, savanna, grasslands, and coastal forests

SIZE
Small to medium: ⅓–¾ in (8–20 mm)

HOST PLANT FAMILIES
Fabaceae, Sapindaceae, Aizoaceae, Rubiaceae, Mimosaceae, Connaraceae, Euphorbiaceae, Caesalpiniaceae,

Until recently this group was treated as a subfamily but has been relegated to the status of tribe. It comprises a vast diversity with 131 genera, 1,380 species, and around 2,043 subspecies described. It is evolutionarily the youngest grouping within Lycaenidae according to molecular studies. Members are distributed globally, although some genera are restricted to certain continents: Africa (*Orachrysops*, *Lepidochrysops*), South America (*Nabokovia*, *Pseudolucia*, *Itylos*), Asia (*Pseudozizeeria*, *Talicada*), and Australia (*Sahulana*, *Neolucia*). Species in *Zizula* and *Leptotes* can be found on every continent, except Antarctica. Some species have a single short, thin tail on the hindwing, but most lack such features. The hindwings are more evenly rounded than in most other Theclinae, which typically have numerous tails. These small, not-very-robust butterflies have a weak flight, even compared with other groups of Lycaenidae.

Caterpillars of some species feed on various plants, some on seeds or pods. Some species have relationships with ants, including mutualism and parasitism. They include some well-studied European species with noteworthy life histories. The larvae of many species associate with ants, which "milk" them. Caterpillars of the European Large Blue *Phengaris arion* initially feed on host plants and are "milked" by ants for their secretions, then taken to the ant nest for protection. Their later pupae mimic the queen and pupate in the ant's nest. This is a risky strategy since some caterpillars are discovered doing this and may be devoured by the ants.

Urticaceae, Ulmaceae, Myrsinaceae, Sterculiaceae, Anacardiaceae, Proteaceae, Boraginaceae, Loranthaceae, Epacridaceae, Santalaceae, Rutaceae, Geraniaceae, Rosaceae, Lamiaceae, Oxalidaceae, Crassulaceae, Polygonaceae, Plumbaginaceae, Gentianaceae, Campanulaceae, Rhamnaceae, and many others

CONSERVATION
The Large Blue *Phengaris arion*, Spanish *Plebejus zullichi*, Italian *Polyommatus humedasae*, and Odd-spot Blue *Turanana taygetica* are categorized as Endangered on the European Red List of Butterflies. The Large Blue became extinct in the UK but has been successfully reintroduced

TOP LEFT | The Red Pierrot *Talicada nyseus* is a tiny butterfly found in the Indo-Malayan region. Its host plants are common garden species from the Crassulaceae family, commonly known as stonecrops.

TOP RIGHT | Large Blue *Phengaris arion*, a species endangered in Europe but extinct in the UK, has became an icon for the conservation of butterflies because of the remarkable work of researchers trying to protect it.

NYMPHALIDAE
BRUSH-FOOTS

The Nymphalidae shows some of the greatest diversity of shapes, patterns, and colors of all the butterfly families. With over 6,800 species described and likely hundreds more yet to be named, it is the most successful group, containing a third of the planet's known butterfly species and popular groups, such as the peacocks, monarchs, postmen, fritillaries, glasswings, ringlets, and many others. All Nymphalids share one character: a reduction in the forelegs. This means it is obvious at first sight that they have only four fully functional walking legs. In males these look like a pair of small brushes, giving rise to their common name, brush-foots. Nymphalids walk only on the four hind legs, which in females also have the sensory role of inspecting host plants for suitability for egg-laying. The antennae have three longitudinal ridges in the middle section on the ventral side.

The morphological diversity of this family is reflected in the 12 subfamilies currently recognized (although academic opinion varies) and 44 traditionally recognized tribes, more numerous than in any other butterfly group. Although Nymphalidae is defined as a family, its taxonomic arrangements and relationships at all levels, from genus to subfamily, have puzzled researchers for many years, resulting in a dynamic and unsettled situation. Here, the latest taxonomic arrangements are followed, which attempt to

balance traditional morphological and ecological groupings with results of recent studies including molecular data.

The Nymphalidae family includes numerous familiar and widespread species—for many people, the butterflies they see most regularly. They have attracted the interest of biologists because of the extraordinary diversity of their life histories and ecological adaptations. The Heliconiinae, for example, has become a model group for studies of tropical evolution, genetics, toxicity, and mimicry. Classic migration studies have centered on the Monarch in the Americas and the Painted Lady in Europe and Africa, whose migrations are well known to the general public.

They occur globally, with diversity peaking in the tropical regions, although they also account for the largest diversity in the Palearctic. Despite some species being widespread, and even migratory, the subfamily Satyrinae has the highest incidence of endemism of all butterfly groups. Many species have ecologies linked to particular plant species, and higher-elevation species are often restricted to narrow ranges, in particular mountain systems in Neotropical and Afrotropical regions, for example. Most Palearctic fauna have received detailed IUCN threat assessments, but only a few of the vast diversity of the tropical members of this group have been similarly assessed, where range-restricted species are likely to be threatened.

Nymphalids are considered to have first arisen on the planet approximately 70 mya. There are various well-preserved fossils of nymphalids, more than from any other family, implying that its members have been successful and diverse throughout recent geological history. Fossils include *Dynamine alexae* (Biblidinae), found preserved in amber, which flew 15–20 mya in the Dominican Republic, as well as numerous species such as *Vanessa amerindica*, *Barbarothea florissanti*, *Chlorippe wilmattae*, *Jupiteria charon*, *Lithodryas styx*, *Prolibythea vagabunda*, and *Prodryas persephone*—which were all discovered within Eocene rock deposits (28–33 million years old) in Colorado, USA.

BELOW | The Clearwing-Satyr *Cithaerias pireta* can be seen gliding at ground level in Neotropical forests.

SNOUTS AND CLUB BEAKS

BELOW | The American Snout *Libytheana carinenta* displaying its long and characteristic labial palps, which give it its common name.

Butterflies in the Libytheinae subfamily can be recognized by two very long, snout-like projections displayed on their heads, which are covered with large, hair-like scales. These are called labial palps. The forewings are also characteristically jagged, and wedge-shaped. However, similar features are found in a few other species in Nymphalidae. Although attempts to study the taxonomy of the group, including comprehensive molecular studies, have been made, there is no consensus on the number of valid species. Approximately 15 species and 23 subspecies are recognized in the genus (with some authors doubling these figures). They are consistently arranged in two genera, *Libytheana* and *Libythea*. Most species have orange-and-brown patterns on the upperside of the wings, except for the iridescent blue *Libythea geoffroyi*; all are pale brown on the underside, resembling dead leaves. Forelegs in females are fully developed, in contrast to most Nymphalids.

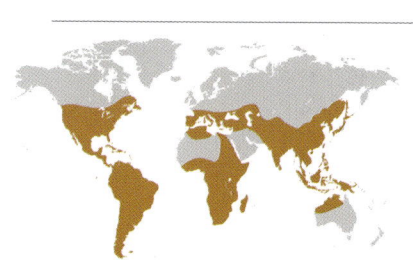

GENERA
Libytheana and *Libythea*

DISTRIBUTION
Mostly tropical in all continents. The genera *Libytheana* is found exclusively in the Americas and *Libythea* in the rest of the world, but not in Antarctica

HABITATS
In gardens, cities, shrublands, near streams, evergreen and tropical forests and forest margins, roads, shady areas, and even

feeding on perspiration from vertebrates. *Libythea* occurs from sea level to 10,000 ft (3,000 m) (Indian *L. myrrha*); American *Libytheana carinenta* up to 7,000 ft (2,100 m)

SIZE
Medium: 1–1⅓ in (25–35 mm)

HOST PLANT FAMILIES
Celtidaceae, Cannabaceae, Ulmaceae, and Rosaceae

ABOVE | Adults of the American Snout *Libytheana carinenta* perching on branches of trees. The patterns on the underside of the wings imitate dead leaves.

RIGHT | Adults of the European Beak or Nettle-tree Butterfly *Libythea celtis* are found in many habitats, including riverbanks and forests, where their host plant *Celtis australis* grows.

CONSERVATION
Madagascar endemic *Libythea cinyras* is categorized as Extinct by the IUCN. No comprehensive threat assessments of other species have been conducted; however, *Libytheana carinenta* and *Libythea laiuus* are categorized as Least Concern by the IUCN. No species in this subfamily are listed in CITES

The biology of the butterflies in this group includes fascinating stories. For example, the Mauritius endemic *Libythea cinyras* is known only from one specimen collected there in 1865 (see p. 41). Because no other specimens have been found since and the Mauritius forests have undergone a dramatic increase in deforestation rates, reducing suitable habitats, the species is now categorized as Extinct. A Madagascan endemic, *Libythea ancoata*, was apparently last captured in 1893, bringing to attention a possible case of extinction on the island.

Species in this subfamily have an unusual distribution, containing both very widespread species, such as *Libytheana carinenta*, and local species, including several island endemics, such as *L. collenettei* (French Polynesia, Marquesas Islands) and *Libytheana fulvescens* (Dominica). Among the butterflies found only in the Americas, there are

ABOVE | European Beak *Libythea celtis* feeding from a flower. The patterns on the underside of its wings resemble dead leaves.

two extinct species, *Libytheana florissanti* and *Libytheana vagabunda*, described from fossils found in the USA.

Migration is a well-recorded phenomenon in Libytheinae. African species *Libythea labdaca* has been recorded on migration in six countries, while *Libytheana carinenta* has been recorded from Mexico to the USA, in Brazil, and in Argentina. Records include up to 1,000 adults at the same time, with estimated numbers of up to 75 million butterflies in an hour. Migration numbers correlate to the amount of rain.

Adults are strong, fast flyers and keep close to the ground; they rest in shade during the hottest hours of the day. The widespread *L. geoffroi* displays aggressive territorial behavior, chasing other butterflies entering its patch while it is basking. Eggs are barrel-shaped with longitudinal ribs, pale cream in color, and laid individually on the inner side of leaves. Caterpillars have a brown head in the early instars; in European *Libythea celtis* later instars become green with black mandibles, resembling larvae in Pieridae. Pupae are elongated and pale green, brown, gray, or black and white with spots. They are suspended from leaves using a secreted silken thread. Flies (tabanids), wasps, and fungi are some of the natural enemies controlling caterpillar populations in the field.

ABOVE | The Club Beak *Libythea myrrha* is found in India and Southeast Asia. Males visit puddles in riverbanks to get essential nutrients.

CLEARWINGS AND GLASSWINGS

The Danainae subfamily is comprised of three historically stable groups that are now recognized as three tribes: Danaini, Tellervini, and Ithomiini, which encompass around 545 species and more than 2,027 subspecies arranged in around 57 genera distributed globally, but mostly found in Asian and Neotropical areas. Together with the Libytheinae they are a sister group to all the other diverse forms of Nymphalidae butterflies.

The Ithomiini are strictly Neotropical. For many years the tribe was considered a subfamily of the brush-foots. However, the tribe's position within the Danainae subfamily is now widely accepted and supported by detailed studies. There are around 396 species and more than 1,542 subspecies described with many more yet to be named. Butterflies in this tribe are characterized by a costal hair pencil in the male's hindwing. The clearwings and glasswings get their common names from the wings of many species being almost completely transparent. However, some display tiger-pattern (black and orange stripes) complexes that are among the most fascinating of all butterflies, and various species are involved in mimicry complexes. Pupae can be shiny silver or gold or simply green or yellow.

Ithomiini are typically inedible for predators because of toxic substances they extract from their host plants. They have been frequent subjects of biogeographical studies and are considered bioindicators of the quality of the environment.

GENERA
Elzunia, Tithorea, Aeria, Athesis, Eutresis, Athyrtis, Paititia, Olyras, Patricia, Melinaea, Methona, Thyridia, Scada, Sais, Forbestra, Mechanitis, Aremfoxia, Epityches, Hyalyris, Napeogenes, Hypothyris, Placidina, Pagyris, Ithomia, Megoleria, Hyposcada, Oleria, Ceratinia, Callithomia, Dircenna, Hyalenna, Episcada, Haenschia, Pteronymia, Velamysta, Godyris, Veladyris, Hypoleria, Brevioleria, Mcclungia, Greta, Heterosais, and *Pseudoscada*

DISTRIBUTION
Neotropical

HABITATS
Forests and forest clearings and edges

SIZE
Small to medium: ⅔–1½ in (15–40 mm)

HOST PLANT FAMILIES
Solanaceae, Apocynaceae, and Gesneriaceae

BELOW | Glasswings, like this *Greta oto*, are popular butterflies among the public because of their "transparent" wings and slow and gracious flight. They are widespread throughout Central and South America.

CONSERVATION
No species have been categorized in the IUCN Red List from this tribe; however, the following species have been comprehensively assessed and are included in the Red Book of Endangered Fauna in Brazil: *Episcada vitrea* and *Scada karschina delicata* are listed as Endangered; *Hyalyris fiammetta, Hyalyris leptalina leptalina, Mcclungia cymo fallens, Melinaea mnasias,* and *Napeogenes rhezia rhezia* are listed as Critically Endangered

OPPOSITE LEFT | The Clearwing *Dircenna klugii* feeds on flowers of shade trees in coffee plantations. Trees of *Inga* sp. aid biodiversity, soil health, and pest control. Adults can also feed on bird excrement.

OPPOSITE RIGHT | An adult of the Tigerwing *Forbestra olivencia*. Life is complicated during its inmmature stages, as they are attacked by Crematogaster ants as well as braconid wasps.

MONARCHS, MILKWEEDS, CROWS, BLUE TIGERS, AND TREE NYMPHS

The cosmopolitan tribe Danaini has around 168 species and 813 subspecies arranged in 14 genera. Milkweed butterflies get their name from ingesting the toxic sap produced by their host plants. These medium- to large-sized butterflies have tough, strong wings that are often colorful with black-marked reds, yellows, oranges, and whites warning of their unpalatable nature. Some species give off a pungent odor. Males are recognized by eversible hair pencils enclosed in pockets in the abdomen. They often have androconial structures on the wings. Various species are involved in mimicry complexes.

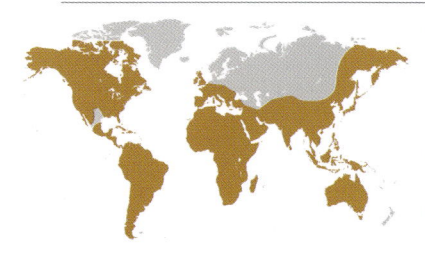

GENERA
Parantica, Miriamica, Ideopsis, Amauris, Tirumala, Danaus, Tiradelphe, Euploea, Idea, Protoploea, Lycorea, Anetia, Archaeolycorea, and *Tellervo*

DISTRIBUTION
All continents except Antarctica

HABITATS
Many primary and secondary habitats, including woodlands, tropical forests, and savanna

SIZE
Medium to large: 1–2¾ in (24–70 mm)

HOST PLANT FAMILIES
Asclepiadaceae, Rosaceae, Apocynaceae, and Moraceae

CONSERVATION
Parantica david, endemic to the Philippines, is categorized in the IUCN Red List as Critically Endangered. The migratory Monarch *Danaus plexippus plexippus* and Asian species *Amauris comorana, Euploea*

Monarchs in the genus *Danaus* are widely used in ecological and biochemical studies. The well-known species *Danaus plexippus* notably undergoes one of the longest and most spectacular butterfly migrations, from central Mexico to northeast USA.

The eggs are barrel-shaped and are laid singly on the underside of leaves. Caterpillars are usually striped, spotted, or hooped, without spines but often with fine hairs. Large species with spectacular patterns include butterflies in the Asian genus *Idea*, some species of which are displayed in butterfly houses around the world.

The Tellervini includes only the single Australasian genus *Tellervo*, with seven species recognized. They are medium-sized, mostly dark, butterflies. Some have white or yellowish eye-like spots or patches on the wings. Eggs are laid on the underside of leaves. Caterpillars bite the stem to forestall defensive reactions from the plant, including suppressing the flow of toxins in the latex, before devouring the leaves. If touched, they drop quickly on a thread. Pupae are green and lack the metallic features of Danaini.

OPPOSITE | The Large Tree Nymph *Idea leuconoe* is a popular and charismatic butterfly bred in butterfly houses around the world because of its large size and slow flight, and also because of its attractive shiny-metallic pupa.

RIGHT | A congregation of Monarchs *Danaus plexippus*, probably the most famous butterfly in the world, not only for its striking patterns but also for its extraordinary migration, one of the most remarkable natural phenomena in the insect world.

albicosta, *Euploea caespes*, *Euploea mitra*, *Euploea tripunctata*, *Ideopsis hewitsonii*, *Parantica kuekenthali*, *Parantica marcia*, *Parantica milagros*, *Parantica schoenigi*, *Parantica sulewattan*, *Parantica timorica*, and *Tiradelphe schneideri* are all categorized as Endangered

EMPERORS, CLIPPERS, GLIDERS, AND SAILORS

The Limenitidinae subfamily has over a thousand species and a similar number of subspecies that are difficult to accurately account for. Species are arranged in 44 genera and seven tribes, with the highest diversity occurring in Africa and Southeast Asia. Only one genus, *Adelpha*, occurs in the Neotropical region and there are four in Australia, including *Neptis*, which is cosmopolitan.

Morphological characters related to male genitalia and host plant preferences have been widely used to group species in Limenitidinae. Adults are charismatic in appearance, fast and strong flyers, and males can be very territorial.

The tribe Adoliadini contains the most species and its members are found across Asia and Africa, although no genera are common to both,

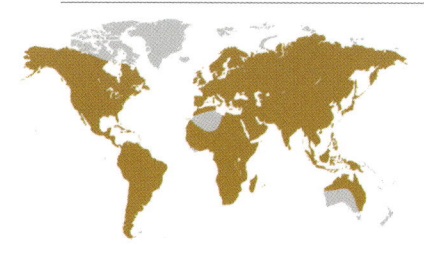

GENERA
Euthalia, Euthaliopsis, Tanaecia, Neurosigma, Abrota, Dophla, Lexias, Bassarona, Catuna, Hamanumida, Euptera, Pseudathyma, Aterica, Pseudargynnis, Cynandra, Euryphura, Euryphaedra, Euphaedra, Harmilla, Euriphene, and *Bebearia* (Adoliadini); *Cymothoe, Harma*, and *Bhagadatta* (Cymothoini); *Limenitis, Sumalia, Moduza, Patsuia, Lamasia, Parasarpa, Tarattia, Auzakia, Litinga, Adelpha, Pandita, Athyma, Kumothales*,

and *Seokia* (Limenitidini); *Cymothoini, Lebadea, Neptis, Pantoporia, Lasippa, Phaedyma*, and *Aldania* (Neptini); *Pseudoneptis* (Pseudoneptini); *Parthenos* (Parthenini); and *Pseudacraea* (Pseudacraeini)

DISTRIBUTION
All continents except Antarctica

HABITATS
Primary forests, secondary forests, forest clearings, and riversides; from sea level

suggesting that the two continents' faunas split a very long time ago. They include large and colorful genera, such as the stunning African genus *Euphaedra*, which alone has over 200 species, and the Asian *Euthalia*, with over 100 species. Caterpillars in this tribe are remarkably decorated and colorful.

The tribe Cymothoini has only three genera (*Harma* is often included in another genera), including African *Cymothoe*, which shows marked sexual dimorphism and has the most species and subspecies across Africa.

BELOW | A caterpillar of the stunning Gaudy Baron *Euthalia lubentina*, a species that inhabits forests in Asia and is legally protected in India.

to 10,000 ft (3,000 m) for Neotropical species

SIZE
Small to large: ⅔–1 ¾ in (17–46 mm)

HOST PLANT FAMILIES
Fabaceae, Rosaceae, Tiliaceae, Annonaceae, Phyllanthaceae, Rubiaceae, Bombacaceae, Boraginaceae, Convolvulaceae, Euphorbiaceae, Sterculiaceae, Ulmaceae, Sapotaceae, and Ochnaceae

CONSERVATION
The African species *Cymothoe teita*, *Neptis katama*, and *Pseudathyma nzoia* are categorized as Endangered in the IUCN Red List. No species in this subfamily are listed in CITES. *Euripus nyctelius* is protected in India

OPPOSITE | In Costa Rica, caterpillars of *Adelpha fessonia* have been observed feeding in a distinctive manner, leaving only the veins of the leaves they consume.

ABOVE | The Common Ceres Forester *Euphaedra phaethusa*. There are nearly 200 recognized species in this genus, making it one of the most diverse groups of diurnal butterflies in Africa.

The tribe Limenitidini includes 14 mostly Asian genera of which many are monotypic, including China's beautiful endemic *Patsuia sinensium*. The only representative of the subfamily in the Neotropical region, the charismatic *Adelpha*, contains species that are hard to separate using adult morphology but can be separated in their distinctive immature stages. Asian and European species in the tribe feed on flowers.

The tribe Neptini has six genera found mostly in Southeast Asia. It includes the widespread genus, *Neptis*, which also occurs in Africa and Europe, but does not occur in the Americas. The sailors, as they are popularly known, have an impressive diversity of over 150 recognized species, which at first sight look very hard to tell apart because of their almost homogeneous black upperside with white stripes. However, species are separable using wing morphology and, in some complex groups, using

characters of the genitalia. Adults are fast flyers found in the canopy of pristine forests, although few species are found in disturbed areas, such as the Common Sailor *Neptis hylas*.

The tribe Parthenini includes only the genus *Parthenos*, known as the clippers, inhabitants of forests in Southeast Asia. Butterflies in this group are widely used in butterfly houses across the world because of their colorful appearance and graceful, gliding flight.

The African tribes Pseudacraeini and Pseudoneptini have one genus each, the complex *Pseudacraea* and *Pseudoneptis*, respectively. Species in *Pseudacraea* are often studied because they mimic species in other subfamilies, such as Heliconiinae (*Acraea*) and Danainae. Also, in certain species such as *Pseudacraea eurytus*, both males and females are mimics, as well as being polymorphic.

LEOPARDS AND YEOMEN

ABOVE | A male Cruiser *Vindula erota*. This species shows a marked dimorphism, with bright green and white females looking very distinctive from the males.

The Heliconiinae subfamily includes some of the most cherished and popular butterflies for public and researchers alike. There are approximately 667 species and more than 1,500 subspecies described. Following recent molecular studies, the subfamily and its various tribes have

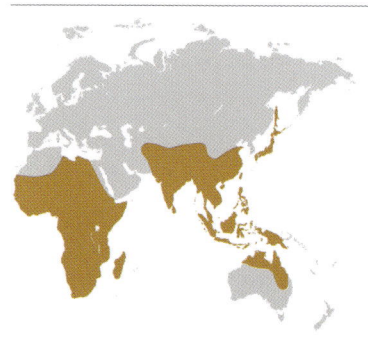

GENERA
Vindula, Lachnoptera, Algia, Algiachroa, Cirrochroa, Terinos, Cupha, Smerina, Vagrans, and *Phalanta*

DISTRIBUTION
East Africa, Madagascar, Asia, and Australia

HABITATS
Mostly forests

SIZE
Medium to large: 1–1¾ in (25–45 mm)

HOST PLANT FAMILIES
Flacourtiaceae, Passifloraceae, and Achariaceae

CONSERVATION
Smerina manoro is a monotypic genus endemic to eastern forests of Madagascar. Being both evolutionarily distinct and endangered, it should be regarded as a conservation priority

had to be substantially redefined to include some genera that had previously been placed elsewhere, such as the genus *Cethosia*. Various shared characters have emerged, including a forewing that is elongated to differing extents in each tribe. Unpalatable species feed on plants that contain cyanogenic glycosides, which are assimilated or processed so as to make the caterpillars and adults poisonous to predators. Butterflies in Heliconiinae have developed bright colors, usually reds, yellows, and oranges, which serve as defensive strategies against more advanced predators such as birds, who learn to avoid eating them.

Four tribes are now recognized, each of which has quite distinct geographical and morphological patterns. Each of these has been redefined by recent molecular studies, so further morphological studies are needed to identify morphological and behavioral commonalities among their genera.

The Vagrantini, with ten genera, are a cosmopolitan (excluding the Americas) and rather contrasting grouping that includes various species with characteristic orange coloration and black marks in the wings. These butterflies are often most active in the morning, when they can be conspicuous and often bask in the sun. Unlike some other tribes, some are fast flyers.

Species in the Indo-Australian genus *Vindula* unusually lay their eggs in dead leaves as a strategy to avoid predation. Many caterpillars of this group are covered in tubercules and setae, a defense strategy warning of unpalatability. Some species feed on flowering plants as adults, but members of other genera also take water from wet soil, puddles, and streams, and some even feed on carrion.

ABOVE | A male of the Malayan Assyrian *Terinos clarissa*, a colorful butterfly found in Southeast Asia. Males have specialized scales and metallic purple forewings.

LEFT | An adult of the Vagrant *Vagrans egista*, a butterfly widely spread across India, South and Southeast Asia and also Australia.

FRITILLARIES AND POLKA DOT

BELOW | The Silver-washed Fritillary *Argynnis paphia* is one of the largest butterflies in Europe. Although listed as not threatened, some populations are a priority because of declines in certain regions.

The Argynnini tribe includes butterflies known as fritillaries, which are most diverse in temperate climates and northern latitudes. Argynnini includes nine genera, over 100 described species, and more than 570 subspecies, previously regarded as a subfamily. Although the forewing is moderately elongated, this is not as pronounced as in other Heliconiinae. Adults often have mottling or spotting, usually arranged in wavy patterns on both the hindwings and forewings, and undulated edges to the hindwings. Males and females are very similar in appearance. Adults feed mostly on flowers. Most species in this tribe are orange, although some are more yellow, brownish, or even black in their base coloration, with patterned ventral sides including silver scales. These are somewhat smaller than other members of this subfamily.

Separation of species is mostly based on wing patterns; however, there is significant variation within populations, making this group challenging to identify for scientists as well as the community of amateur taxonomists in the Palearctic region.

GENERA
Euptoieta, Pardopsis, Yramea, Boloria, Issoria, Brenthis, Argynnis, Speyeria, and *Fabriciana*

DISTRIBUTION
Mostly temperate zones in the Holarctic region. Some species in the Neotropics, Afrotropics, Asia, and Australia

HABITATS
Temperate montane and submontane forests, secondary forests, and grasslands

SIZE
Small to large: ⅔–1¾ in (17–45 mm)

HOST PLANT FAMILIES
Ericaceae, Passifloraceae, Salicaceae, Saxifragaceae, and Violaceae

CONSERVATION
The Scandinavian Dusky-winged Fritillary *Boloria improba* is listed as Endangered in the European Red List of Butterflies

Although this tribe is most diverse in the Holarctic region, there are some notable exceptions—for example, members of the genera *Yramea* and *Euptoieta* are found in temperate regions of southern South America, such as Patagonia and the Andes. The Polka Dot *Pardopsis punctatissima* is a monotypic bright yellow butterfly with black spots, which is widely distributed in the lowlands and mountains of Africa and Madagascar. Finally, the genus *Issoria* includes members in the Mediterranean region and tropical latitudes of Africa.

LEFT | Fritillaries are popular with the general public in North America because of their charismatic looks. The Crown Fritillary *Speyeria coronis* is found in grassland and oak woodlands.

BELOW | A male adult of the Polka Dot *Pardopsis punctatissima*, found in Afromontane forests. Its flight is low and weak and some species in the Lycaenidae seem to mimic it. It is locally abundant.

POSTMAN AND GULF FRITILLARIES

RIGHT | The Clysonymus Longwing *Heliconius clysonymus* is found in central and South America.

Members of Heliconiini are among the most conspicuous and emblematic Neotropical butterflies. They include eight exclusively Neotropical genera, 77 species, and more than 457 recognized subspecies. Species in this group can be found from the lowlands up to the highest elevations, in paramo habitats, primary and secondary forests, other secondary habitats, and flowering gardens. Adults exclusively feed on nectar and are pollinators. *Heliconius* butterflies have a highly elongated forewing, giving them a distinctive oblong shape overall, long antennae, and large eyes. Many species are toxic to predators (or are mimics of co-occurring species that are toxic). To draw attention to this toxicity, they are often slow moving and showy. Being associated with flowers of the understory, they are easily observed and have been widely collected.

Eggs may be laid in groups or singly. The caterpillars have many spines, which act as a warning and irritant to predators; handling by humans will often bring on a rash. Pupae may be spined or smooth and are suspended from a branch or held parallel to it. Species in this group are often reared in captivity and butterfly houses, leading to a good understanding of their life histories. The

GENERA
Philaethria, Dryadula, Dryas, Podotricha, Agraulis, Dione, Eueides, Heliconius, and *Neruda*

DISTRIBUTION
Exclusively Neotropical and southern USA

HABITATS
All forested habitats and secondary growth, from sea level up to 6,000 ft (1,800 m)

SIZE
Medium to large: 1¼–2 in (30–50 mm)

HOST PLANT FAMILIES
Plants from families Passifloraceae and Turneraceae

CONSERVATION
Heliconius nattereri is categorized as Critically Endangered on the IUCN Red List and Endangered in the Brazilian Book of Endangered Fauna. No other species in this tribe have been listed as threatened. *Heliconius erato cyrbia* was introduced in 2016 to the Cook Islands to control Red Passionfruit (*Passiflora rubra*)

RIGHT | The
Flame *Dryas iulia*
is found across
south Florida,
Central America,
the Caribbean
islands, and South
America to Brazil.
Its flight is slow
and elegant.

genus *Heliconius* has become one of the best-studied groups of insects. They are regarded as a model group for understanding evolutionary and molecular biology, and biogeography. Evolutionary studies have historically focused on coevolution with host plants and the mimicry rings that these species are involved in. Their place as a model study group is long-standing: Henry Bates originally conceived mimicry based on his study of these butterflies in Amazonia in the 1800s.

ABOVE RIGHT | The vibrant orange color of the Banded Orange *Dryadula phaetusa* adults serves as a warning to predators about the unpalatability of this species.

RIGHT | The Scarce Bamboo *Philaethria dido* is widespread in the Neotropics. Its bright-green coloration often leads to it being confused with Malachite *Siproeta stelenes*.

ACRAEA, ACTINOTE, AND LACEWINGS

The Acraeini is a diverse group of colorful tropical butterflies that have an elongated and often pointed forewing and waved edges to the hindwings. The Acraeini are mostly African but have one major radiation in South America containing five genera. The antennae of this group lack dense scaling, and the clubs at the end of the antennae are well-developed. The wing cells are closed by tubular veins. They often fly slowly, which shows off their bright warning marks to predators. Most species are brightly colored, especially with reds, yellows, and oranges, which serve as a

RIGHT | The Red Lacewing *Cethosia biblis* shows not only striking patterns on the underside of the wings, but also vibrant colors on the upperside, making it among the favourite species found in butterfly houses. This species is found in forests in Southeast Asia.

GENERA
Cethosia, Acraea, Bematistes, Telchinia, and *Actinote*

DISTRIBUTION
Tropical regions of Central and South America, Africa, Asia, and north Australasia

HABITATS
Forests, paths, secondary growth forest, bushes along rivers, and Andean cloud forest (*Actinote*) at elevations up to 9,000 ft (2,700 m)

SIZE
Medium: ¾–1½ in (22–38 mm)

HOST PLANT FAMILIES
Urticaceae, Passifloraceae, Asteraceae, Fabaceae, Flacourtiaceae, Violaceae, and Turneraceae

CONSERVATION
Brazilian endemic *Actinote zikani* is listed as Critically Endangered in the Brazilian Book of Endangered Fauna. The Ethiopian species *Telchinia guichardi* is reported in

warning of toxicity. There are three large genera recognized, one occurring in each of the tropical continents.

The hyperdiverse genus *Acraea* (into which *Bematistes* and *Telchinia* are often lumped) includes most African species, together with a few Asian and Australasian representatives. Numerous subgroups have been identified, and these may merit recognition as genera. This group supports some of the best-studied examples of mimicry complexes in the Asian and African regions. The lacewings *Cethosia* were recently placed in this tribe following molecular studies. They are largely restricted to tropical Southeast Asia, with a handful of species ranging into northern Australia. They often have red or orange markings toward the center of the body (sometimes purple), with black markings, white lines, or bands and marked serrations or waves on the hindwings. Individuals of some species of South American *Actinote* congregate in groups of hundreds along streams and at puddles, where they take nutrients.

LEFT | The genus *Actinote*, found only in the Neotropical region, has over 50 recognized species with more yet to be described. Many species exhibit Müllerian mimicry, sharing similar warning color patterns on the wings with species that are inedible to predators.

BELOW | The Tawny Coster *Acraea terpsicore* is widely distributed across India. Caterpillars are known to feed on Passion Vines *Passiflora*, although recently there are records of females laying eggs in the wild flower *Turnera subulata*.

specialized literature as possibly extinct due to the draining of marshes where it was found more than 70 years ago. No IUCN assessment has been conducted. *Telchinia induna salmontana* is nationally listed in South Africa as Endangered

NINJA, CONSTABLES, TABBIES, AND POPINJAYS

This is a small subfamily of butterflies found in the Indo-Australian region (but not in Australia), which comprises four genera and seven recognized species. The taxonomic relations of this group have been controversial, and some have been placed in different subfamilies or tribes. Butterflies in this group are mostly restricted to tropical forests and have intricate and often extensive mottled patterns, which usually extend across both the forewings and the hindwings. Females are similar but paler and slightly bigger than males, showing various forms in some species.

One of the more distinctive members of this subfamily is the monotypic Ninja butterfly

GENERA
Amnosia, Pseudergolis, Stibochiona, and Dichorragia

DISTRIBUTION
Continental Southeast Asia, from India to China, the Malay peninsula, and islands of tropical Asia to Borneo

HABITATS
Primary and mature secondary forests, mostly in the foothills and premontane elevations at 1,300–5,000 ft

(400–1,500 m). Stibochiona are found in the understory of forests or hilltopping on stunted growth, rocks, and grasses on mountaintops

SIZE
Medium to large: ¾–1⅔ in (22–40 mm)

HOST PLANT FAMILIES
Urticaceae (Amnosia, Stibochiona, Pseudergolis), Anacardiaceae, Meliosmaceae, Burseraceae, Sabiaceae (Dichorragia, Pseudergolis), and Moraceae (Stibochiona)

Amnosia decora, which encompasses six subspecies. It is found on mainland Asia and tropical Asian islands at low elevations. This species is often observed on tree trunks, where it feeds on sap. Members of genus *Pseudergolis* are known as tabbies because of the bright orange patterns on the upperside of the wings, although they are dark brown on the underside. There are two species and five subspecies recognized including *Pseudergolis avesta*, which is endemic to Sulawesi and *Pseudergolis wedah*, which is found in the Himalayas in India.

The genus *Stibochiona*, known collectively as popinjays, has just two species and six subspecies with purplish, bluish, or greenish base colorations in the wings. Their eggs are globular with vertical ridges, and caterpillars are dark greenish-brown, mostly smooth, and have large horns on the head and tiny yellow spines only near the tail. Pupae resemble dry leaves. The genus *Dichorragia*, known as constables, also has just two species with 16 subspecies recognized. Caterpillars of *Dichorragia nesimachus* are green and smooth, and pupae are reddish-brown.

OPPOSITE | The Constable *Dichorragia nesimachus* is a stunning butterfly found in Asia that displays great variation across the region. Its caterpillars are peculiar, displaying giant, thick, horn-like appendages.

RIGHT | The Tabby butterfly *Pseudergolis wedah* is found in South Asia. Caterpillars of this species feed on plants from the Urticaceae family.

BELOW RIGHT | Males of the Popinjay *Stibochiona nicea* have velvet blue wings, while females have distinctive green coloration.

CONSERVATION
No species in this subfamily have been assessed for the IUCN Red List and no species are listed by CITES. However, *Dichorragia nesimachus* has been listed as a species recorded for protection in South Korea since 1981. The subspecies *Dichorragia nesimachus pelurius* is endemic to Sulawesi

EMPERORS

The subfamily Apaturinae is distributed almost globally, with most genera found in Eurasia and Southeast Asia, except for *Apaturina*, which extends to the Australasian region. *Apaturopsis* is the only genus in Africa, and *Doxocopa* and *Asterocampa* are distributed in the Americas only (South America and North America, respectively).

The adults are generally medium-sized, and males have colorful, sometimes iridescent, patterns on the upperside with purple, or black-and-white colors, although cryptic in the underside of the wings. Within the subfamily, there are 21 recognized genera, 96 species, and over 223 subspecies described, with hundreds of names for

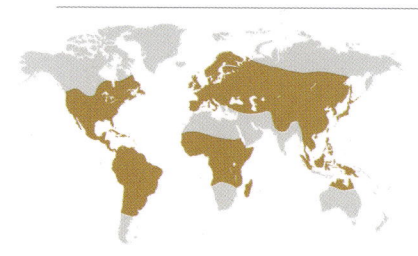

GENERA
Apatura, Mimathyma, Chitoria, Dilipa, Euapatura, Eulaceura, Rohana, Euripus, Helcyra, Herona, Hestina, Hestinalis, Sasakia, Sephisa, Apaturina, Apaturopsis, Timelaea, Thaleropis, Asterocampa, Doxocopa, and *Lelecella*

DISTRIBUTION
Worldwide, except Antarctica

HABITATS
Tropical forests, treetops, moisture on sunny paths and riversides

SIZE
Medium: 1–1 ⅓ in (25–35 mm)

HOST PLANT FAMILIES
Salicaceae (*Apatura*), Ulmaceae (*Doxocopa*), Cannabaceae, Betulaceae, and Fagaceae

CONSERVATION
Only species found in Europe have been subject to conservation assessments by the IUCN. For example, although regarded as Near Threatened or of Least Concern under European statute, *Apatura iris* is protected

OPPOSITE | The striking *Sasakia charonda* is the national butterfly of Japan; it is commonly known as the Japanese Emperor.

ABOVE | The extraordinary iridescence and colors of the Purple Emperor *Apatura iris* are produced by microscopic structures in its wings. Although widespread in Europe and temperate Asia, this species is in decline in the UK.

RIGHT | The Turquoise Emperor *Doxocopa laurentia* is found in forests in South America. The taxonomy of the genus remains unresolved.

under the Wildlife and Countryside Act in the UK. Other nationally protected species include *Euripus nyctelius* in India. *Doxocopa zalmunna* is listed as Critically Endangered (and possibly Extinct) in the Red Book of Endangered Fauna of Brazil. No species in this subfamily are listed in CITES

aberrations and variations. The subfamily is defined by unique morphological characters in the male genitalia (elongated phallus) as well as the arrangement of veins in the forewings. Although recent molecular studies have been conducted, there is no agreed consensus on the taxonomy of the group. Also, because various genera and species have not been studied in detail, there are various disagreements in species limits, and there are certainly many undescribed taxa in the group. This subfamily includes an extinct fossil species, Colorado's *Doxocopa wilmattae*, which was described in 1907.

Adults are strong flyers found high in the canopy in well-wooded forests; some are rarely seen unless in bright sun or attracted to baited traps using rotting fruits or carrion. Also, males visit puddles and wet ground along paths and riversides and feed at various organic sources. The South American genus *Doxocopa*—which contains the most species—and spectacular Asian genus *Sasakia* include very attractive iridescent butterflies; they are strong and agile forest flyers, and the males are often attracted to salts on riversides. The females of *Doxocopa* and Asian genus *Euripus* display different forms, mimic other species, and are often confused with species in other genera or even other subfamilies. Females of some species can be larger. The colorful European Purple Emperor *Apatura iris* is a rare species that feeds on sweet substances (honeydew) taken from greenfly (aphids) and also feeds on tree sap. Other species rarely found in the field and in collections include some in the Southeast Asian genus *Helcyra*.

Eggs in the well-studied North American genus *Asterocampa* are dome-shaped, pale yellow, or

white, and ornamented with vertical ridges; caterpillars are green with short lateral spines, and some species display a pair of black horns. Caterpillars of some species in the African genus *Apaturopsis* can be gregarious and are found in large blocks of broadleaved woodland or in dense scrub. This genus includes Madagascan endemic *Apaturopsis kilusa* and *A. paulianii*. There is very little information available about the immature stages of most species in this subfamily.

ABOVE | The stunning courtier butterfly *Sephisa daimio* was discovered for the first time in Taiwan over 100 years ago.

OPPOSITE | A male of North American Hackberry Emperor *Asterocampa celtis* feeding on dung or carrion. They also visit riverbanks to extract important nutrients and minerals.

CRACKERS, BANNERS, JOKERS, AND CASTORS

This subfamily contains perhaps some of the most colorful and attractive medium-sized butterflies, and has one of the most diverse distributions of all butterfly subfamilies. There are currently 307 species and 632 subspecies, currently organized into seven tribes. Members of this subfamily are recognized from certain characters (some shared with other groups) such as the swollen bases to the forewing veins and production of a "frass chain" during the caterpillar stage, which assists them in repelling attacks from predators such as ants. In the early 2000s, Biblidinae was raised from being a tribe in the Limenitidinae subfamily.

Not every species is brightly colored—some, such as *Vila* and *Neptidopsis*, are dark brown or black—yet Biblidinae gets substantial attention

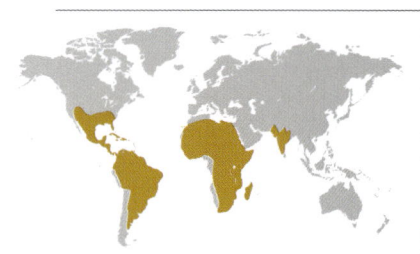

GENERA
Sevenia, Eunica, Cybdelis, Catonephele, Myscelia, Nessaea (Epicaliini); *Dynamine* (Eubagini), *Biblis, Archimestra, Laringa, Ariadne, Byblia, Eurytela, Mesoxantha, Mestra, Neptidopsis,* and *Vila* (Biblidini); *Hamadryas, Batesia, Ectima,* and *Panacea* (Ageroniini)

DISTRIBUTION
Tropical habitats in Central and South America, Africa, and India, mostly in lowlands from sea level up to 7,000 ft (2,100 m).

HABITATS
Pristine forests, secondary forests, humid tropical forests, forest edges, true savanna, disturbed areas, and roadsides

SIZE
Small to large: ⅔–1¾ in (16–45 mm)

HOST PLANT FAMILIES
Euphorbiaceae (most), Lauraceae (*Catonephele*), Arecaceae (*Hamadryas*),

LEFT | The underside of the Neotropical *Nessaea aglaura* helps it to blend in with foliage in Neotropical forests. On the upperside, this butterfly is black with bright blue marks on the forewings and orange in the hindwings.

OPPOSITE | Butterflies in the genus *Catonephele* are found from Mexico through Central and South America. Males have bright orange wings covered in black velvet scales. Females are black and white. These butterflies are attracted to traps with baits.

from collectors, photographers, and scientists alike, resulting in the vast number of specimens in collections and available data. Therefore, consensus in the taxonomic arrangements in this group is of wider interest and debate, and yet the taxonomy is still to be decided.

Most species in Biblidinae do not show a marked dimorphism between males and females,

---

Sapindaceae, and Ulmaceae (*Haematera, Epiphile*)

CONSERVATION

Catagramma hydarnis is listed as Critically Endangered and *Hamadryas velutina browni* as Endangered in the Red Book of Endangered Fauna of Brazil. *Sevenia amazoula* and *S. madagascariensis* are only found in Madagascar and are potentially threatened due to the current alarming rates of deforestation and habitat loss there

although females have rounder wings and are slightly bigger and paler in color.

The tribe Epicaliini includes 67 species and over 100 subspecies, distributed solely in the Neotropics, although, until recently, the African genus *Sevenia* was placed in this tribe. It includes species displaying strong dimorphism, such as in the Neotropical *Catonephele*, in which females are black and males are bright orange. Other interesting species include some with migratory habits, such as African *Sevenia amulia*. Adults feed on rotting fruit and tree sap and can be found near rivers and perching on rocks.

Tribe Eubagini includes solely the genus *Dynamine*, whose caterpillars resemble slugs with tiny spines; they are ferocious eaters of young flowers. *Dynamine* contains the largest number of species and subspecies in Biblidinae (40), although

it is likely that there are more species to describe.

The tribe Biblidini includes the genera *Neptidopsis*, *Mesoxantha*, and *Eurytela*, inhabitants of Africa; *Laringa* which is found only in Asia; and *Byblia* and *Ariadne*, which occur in both continents. All other genera are found in Central and South America. Some species may be very common, such as *Biblis hyperia*. Others, although widely distributed, can be rare locally—for example, *Mestra amymone* in Costa Rica—or only found in certain places—*Archimestra teleboas* is found only on the Caribbean island of Hispaniola and *Eurytela narinda* only on Mauritius. Males and females in this tribe do not show many differences in their wing patterns or colors, although females can be larger and paler. Males in *Mesoxantha* are territorial and strong flyers. African species of

TOP | The stunning *Batesia hypochlora* inhabits lowland rainforests in Colombia, Ecuador, Peru, Brazil, and Bolivia. Adults are frequently observed gliding slowly on sunny days.

ABOVE | The Joker *Byblia ilithyia*. Males and females are almost indistinguishable to the general observer. Females lay eggs singly on the undersurface of leaves of the Indian stinging nettle *Tragia plukenetii*, its larval hostplant.

ABOVE | Crackers in the genus *Hamadryas* exhibit a wide variety of wing colors and markings, making it difficult to distinguish between species and subspecies. These butterflies produce an unusual clicking noise when flying, and are relatively abundant in human-modified areas.

RIGHT | A male of the Common Castor *Ariadne merione*, a widespread butterfly in India and South and Southeast Asia.

Byblia have long spikes on their eggs, which are laid individually under leaves. Caterpillars in Neotropical *Biblis* have spikes and are dark in color.

The tribe Ageroniini includes some of the most spectacular genera from the tropics of South America, including the stunning Amazonian species *Batesia hypochlora*, the crackers in *Hamadryas* which make a "cracking" noise, and the recently recorded migrant Amazonian species *Panaceas prola*. Adults feed on rotting fruit on the ground or in the canopy and are easily attracted to bait.

EIGHTY-NINES

The tribe Callicorini includes some of the most colorful Neotropical species, including the peculiar Eighty-nine, Eighty-eight, and Sixty-eight butterflies in the genus *Callicore*, which display numeric-like patterns on the ventral side of the hindwings. This genus also includes restricted-range species—for example, *Callicore ines*, endemic to pristine jungles in Colombia and *Callicore hydarnis*, endemic to Brazil. Caterpillars are also interesting in this genus as they have a scoli (a long, horn-like structure with tiny, scattered projections) attached to the head, which differ in detail between species. There are 80 species and more than 229 subspecies arranged in between nine and 12 genera in this tribe, including the Amazonian *Antigonis* and the recently described endemic *Archaeogramma* from

LEFT | The stunning Eighty-eight *Callicore texa* is a Neotropical species known for the numerical-like patterns on the underside of its hindwings.

GENERA
Antigonis, Archaeogramma, Callicore, Catagramma (including *Paulogramma*), *Haematera, Perisama* (including *Mesotaenia*), *Diaethria* (including *Catacore*) *Orophila, Sea* (Callicorini); *Asterope, Bolboneura, Callicorina, Epiphile, Lucinia, Nica, Peria, Pyrrhogyra,* and *Temenis* (Epiphilini)

DISTRIBUTION
Central and South America, and the Caribbean

HABITATS
Primary and humid tropical forests, cloud forests, forest edges, and riverbanks at low to middle elevations up to 13,000 ft (4,000 m).

SIZE
Small to large: ⅔–1¾ in (16–45 mm)

HOST PLANT FAMILIES
Euphorbiaceae (most), Lauraceae (*Catonephele*), Sapindaceae, and Ulmaceae (*Diaethria, Haematera, Epiphile*)

the Tepuis in Venezuela. The brightly colored genus *Perisama* has the highest diversity in the subfamily with 40 species and more than 80 subspecies. Recent molecular studies have included the related genera *Mesotaenia* and *Orophila* in *Perisama*. Host plants and immature stages are largely known for Andean species in this tribe.

The Epiphilini is another exclusively Neotropical tribe and includes several uncommon monotypic genera and various brightly colored genera, including *Asterope*, an inhabitant of undisturbed forests in the Amazon, which is iridescent on the upperside and striking in the underside of the wings. In contrast to their pretty appearance, species in the genus *Temenis* feed on excrement. Adults in the genus *Epiphile* feed on tree sap and are attracted to rotting fruits in the forest. Caterpillars in this tribe make frass chains that offer protection with a physical barrier created from the environment to deter predators.

LEFT | An adult of *Perisama humboldtii*, a Neotropical species. The bright yellow underside of the wings contrasts with the upperside which is black with shades of metallic blue.

BELOW | A spectacular male of Neotropical species *Epiphile orea*. Females of various species in this genus remain unknown to science.

CONSERVATION
No species in Biblidinae have been categorized as threatened in the IUCN Red List. However, *Catagramma hydarnis* is listed as Critically Endangered and *Hamadryas velutina browni* as Endangered in the Red Book of Endangered Fauna of Brazil. No species are included on the CITES list

MAPS AND DAGGERWINGS

The three genera, 50 species, and more than 130 subspecies in this small subfamily appear not to have close relationships with other subfamilies, and Cyrestinae has not long been raised as a subfamily, supported by molecular studies. They are charismatic butterflies with a tail-like feature in the hindwings, and are often confused with Papilionids, although they are smaller in size. This group includes the beautiful *Cyrestis nivea*, which has patterns of black lines on a

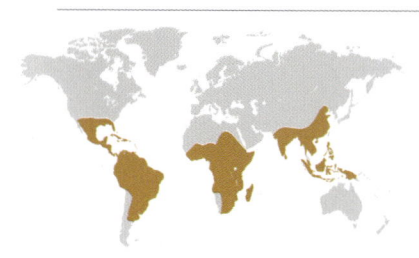

GENERA
Cyrestis, Chersonesia, and *Marpesia*

DISTRIBUTION
Southeast Asia and Neotropics; one species in Africa

HABITATS
Tropical evergreen forests, forest paths, and clearings, from sea level up to 3,200 ft (1,000 m); some species of *Marpesia* can be found up to 6,500 ft (2,000 m).

SIZE
Medium to large: 1¼–1½ in (30–40 mm)

HOST PLANT FAMILIES
Moraceae and Dilleniaceae

CONSERVATION
Cyrestis camillus, Cyrestis nivea, Cyrestis themire, Chersonesia excellens, and *Chersonesia intermedia* are categorized on the IUCN the Red List as Least Concern. No species in this subfamily are included on the CITES checklist. The recently

white ground color that inspired its common name of map butterfly.

Species in the genus *Chersonesia* are often orange with black lines and are inhabitants of the Indo-Malayan region. Although grouped in the same subfamily, it is reported that they differ from *Cyrestis* in wing venation (observable when removing the scales) and having shorter palps. The genus *Marpesia*, fast flyers in forests in Central and South America, includes slightly larger butterflies with colorful as well as dull tones in the wings. Males in Cyrestinae congregate in large numbers near streams or muddy puddles. Females are rarely seen in the field or in collections.

Males and females in most Asian species are similar in appearance; however, males have much darker coloration and triangular-shaped forewings, while those of females are rounder. In *Marpesia*, dimorphism is marked among males and females. The Oriental Map butterfly *Cyrestis thyodamas* displays two distinctive seasonal forms that can confuse amateur taxonomists.

Caterpillars are elongated and have large, stiff processes (as horns or spikes) on the body as well as on the head, as in *Chersonesia peraka*; some have white spots, as in *C. themire*. Eggs in *Marpesia* are white and are laid individually, while the caterpillars can be brightly colored; it is reported that they can be attacked by parasitoid wasps.

described *Marpesia pantepuiana* is endemic to the isolated and unique tabletop mountains called tepuis in the Pantepuis region of Venezuela

OPPOSITE | The Common Maplet *Chersonesia risa*, a locally common species found in montane forests in Malaysia. Males are territorial and are often observed resting on the underside of leaves with their wings spread.

TOP | The Straight Line Map-wing *Cyrestis nivea*, found across montane forests in Southeast Asia. It shows great variation in the map-like patterns on its wings.

ABOVE | Butterflies in the Neotropical genus *Marpesia* are commonly known as daggerwings. They resemble swallowtails due to the similar tail-like extensions to the hindwings.

BRUSH-FOOTS, LEAFS, ADMIRALS, TORTOISESHELLS, COMMAS, AND PANSIES

This is the largest and most speciose subfamily in Nymphalidae with approximately 521 species and 1,459 subspecies, arranged in 57 genera. This group includes butterflies in all colors, sizes, and shapes, which inhabit almost every habitat and continent on the planet, except Antarctica. Among genera in this subfamily are some of the most widely distributed, yet also the most taxonomically challenging—for example, *Aglais*, *Nymphalis*, *Melitaea*, *Hypolimnas*, and *Junonia*. Males are powerful flyers in the forest, colorful in the upperside, and with cryptic patterns in the underside of the wings. Adults visit flowers occasionally but feed also on carrion and fermented fruits. Eggs are barrel-shaped and veined in the surface, and with few exceptions, such as in

RIGHT | A female of the Mimic *Hypolimnas missipus.* This iconic species is well known because of its migration and widespread occurence in almost every continent. It also displays strong dimorphism with bright orange females mimicking unpalatable species.

GENERA
Baeotus, Colobura, Historis, Pycina, Smyrna, and *Tigridia* (Coeini); *Junonia, Precis, Protogoniomorpha, Salamis,* and *Yoma* (Junonini); *Catacroptera, Doleschallia, Hypolimnas, Kallima, Malika,* and *Rhinopalpa* (Kallimini); *Kallimoides* (Kallimoidini); *Aglais, Antanartia, Araschnia, Hypanartia, Kaniska, Mynes, Nymphalis, Polygonia, Symbrenthia,* and *Vanessa* (Nymphalini); *Vanessula* (Vanessulini), *Anartia,* *Metamorpha, Napeocles,* and *Siproeta* (Victorini)

DISTRIBUTION
All continents, except Antarctica

HABITATS
Hardwood boreal and tropical forests, forest openings, savanna, and along streams

SIZE
Small to large: ¾–2 in (19–48 mm)

Australian genus *Mynes*, they are laid individually in the host plant. Caterpillars have rows of scoli on the body and are dark in color; they are gregarious during the first stages of life.

The subfamily includes most of the taxa whose taxonomic placement is uncertain. Host plant and immature stages have provided characters that are diagnostic, although some are shared with other subfamilies. Recent detailed molecular and morphological studies have brought more clarity to the relationships among the group, but more work remains to be done.

TOP | The Mother-of-Pearl *Protogoniomorpha parhassus*, a species found in forests in central and southern Africa. Because of its beauty it is a popular breeding species in butterfly houses around the world.

RIGHT | The African *Precis octavia* is a remarkable species exhibiting seasonal dimorphism, meaning that there are two distinct forms depending on the season. This causes confusion to unaware naturalists and collectors.

HOST PLANT FAMILIES

A wide range of plants from families Acanthaceae, Amaranthaceae, Convolvulaceae, Lamiaceae, Malvaceae, Plantaginaceae, Polygonaceae, Rubiaceae, Scrophulariaceae, Ulmaceae, Urticaceae, and Verbenaceae

CONSERVATION

Despite the wide diversity in this subfamily, only a few species in the genus *Junonia* have been categorized on the IUCN Red List as Vulnerable. *Aglais urticae* is listed as low conservation priority in the UK, but it is listed as a species of concern

The tribe Junonini, with five genera, is mostly Afrotropical, and includes *Precis*, *Protogoniomorpha*, *Salamis*, the cosmopolitan genus *Junonia*—commonly called buckeyes or pansies—and Indonesian *Yoma*. Mother-of-pearls include the endemics *Salamis humbloti*, found only in Madagascar, and *S. anteva* on the Comoro Islands. The species *Precis octavia* shows such remarkable seasonal dimorphism that it can deceive both amateurs and experts; the two forms were described and have been displayed as separated species.

The tribe Kallimini includes six genera of medium-sized butterflies distributed in Africa, such as *Catacroptera cloanthe* and more recently described genera *Malika*; in Asia (*Doleschallia*, *Kallima*, *Rhinopalpa*); or are cosmopolitan, as in *Hypolimnas*, which includes the remarkable sexually dimorphic species *Hypolimnas missipus*, which is distributed virtually all over the world.

The tribes Kallimoidini and Vanessulini each include monotypic genera, with African Leaf butterfly *Kallimoides rumia* and Lady's Maids *Vanessula milca* exclusively found in Africa.

These arrangements were proposed originally based on morphology but are now supported by recent molecular studies. However, further research is needed.

Perhaps the most famous representative of the tribe Nymphalini is the Painted Lady *Vanessa cardui*, which has one of the widest geographical distributions in the world (worldwide except South America) and is the protagonist of one of the most remarkable phenomena of long-distance migration, from Africa to Europe. This tribe comprises ten genera of butterflies, of which five are distributed globally alongside Afrotropical *Antanartia*, Asian *Symbrenthia* and *Araschnia*, Neotropical *Hypanartia*, and Australian *Mynes*. Butterflies in the genus *Polygonia* are powerful flyers and some species are migratory with distinctive marks on the underside of their wings resembling a small C or "comma", giving them the common name of commas.

The tribe Victorini is exclusively Neotropical and includes four genera including common species found in human-made habitats. However, *Napeocles* is uncommon and found in forests.

OPPOSITE LEFT | Wet season form of the African Mother-of-Pearl *Protogoniomorpha parhassus*. It is a widespread species found at elevations from sea level up to 7,875 ft (2,400 m).

OPPOSITE RIGHT | A female in dry season of the Pirate *Catacroptera cloanthe*, a species found in grasslands and savannas in Africa in elevations between 985 and 8,200 ft (300–2,500 m).

RIGHT | The Painted Lady or Cosmopolitan *Vanessa cardui*, an iconic species famed for its remarkable migration from Africa to Europe. It is one of the most widespread species in the world, found in virtually every continent except Antarctica and South America.

BELOW LEFT | The Common Jester *Symbrenthia lilaea* is common in montane forests in India and South Asia. Its caterpillars feed on various species of *Boehmeria* and are gregarious.

BELOW RIGHT | Species in the genus *Kallimoides* are known as African leaves because when their wings are closed, they display dark brown patterns resembling a dry leaf.

CHECKERSPOTS AND CRESCENTS

The tribe Melitaeini used to be considered a subfamily. However, recent molecular works have shown its close relationship with other tribes in the Nymphalinae subfamily. It is probably one of the less-studied groups of brush-foots, and there are still unknown females, host plants, and immature stages in various species, such as in *Anthanassa* and *Microtia*. With 24 accepted genera, the Eurasian genus *Melitaea* contains the richest species diversity with around 280 species and more than 880 subspecies described. Butterflies in this group are medium-sized and are consistently patterned with black markings as lines and/or spots on an orange ground color on the upperside of the wings; the undersides are also patterned. Most Palearctic and Nearctic species are patterned like this, with exceptions including the North American Baltimore Checkerspot (*Euphydryas phaeton*) and the European Cynthia's Fritillary (*Euphydryas Cynthia*).

A flat and well-developed tip of the antenna is a common character among species in this tribe. Below tribe level, structure of the genitalia (in males and females) is well-defined.

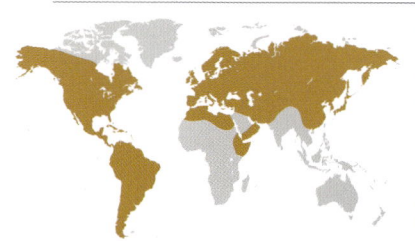

GENERA
Anthanassa, Antillea, Atlantea, Castilia, Chlosyne, Dagon, Dymasia, Eresia, Euphydryas, Gnathotriche, Higginsius, Janatella, Mazia, Melitaea, Microtia, Ortilia, Phyciodes, Phystis, Poladryas, Tegosa, Telenassa, Texola, and *Tisona*

DISTRIBUTION
Holarctic, Neotropical, and North Africa

HABITATS
Woodlands and woodland clearings and edges, shrubs, rocky riparian slopes, open areas, and pastures, from sea level up to elevations of 4,600 ft (1,400 m) in Neotropical species

SIZE
Small to medium: ½–1¼ in (13–30 mm)

HOST PLANT FAMILIES
Asteracea, Acanthaceae, Amaranthaceae, Poaceae, Plantaginaceae, and Caprifoliaceae

Eggs are laid in clusters on leaves, and there are reports that Neotropical *Chlosyne* eggs are already mature before they are laid. Caterpillars in *Anthanassa* are black with spines and live communally in a web of silk; when they are more developed, they become solitary. It is thought that chemicals in *Helianthus* plants (Asteracea) might be ingested by caterpillars and adults making them unpalatable; however, further studies are needed to investigate this.

CONSERVATION
No species in this tribe are on the IUCN Red List or included on the CITES list. However, this group includes restricted-range and rare species, such as *Chlosyne gorgone gorgone*, which has been reported as potentially extinct in the eastern USA. Brazilian *Eresia erysice erysice* is listed as Critically Endangered in the Red Book of Endangered Fauna of Brazil. The European Red List categorizes *Melitaea aurelia* as Near Threatened

ABOVE | A male of Cynthia's Fritillary *Euphydryas cynthia*, a butterfly found in high elevations in the Holarctic region. This species shows a strong dimorphism where females are mostly orange with black lines and marks.

OPPOSITE | The Heath Fritillary *Melitaea athalia*, a European species. Adults are restricted to some specialized habitats, hence it is a conservation priority and is protected in some countries like the UK.

FREAKS

Calinaginae is the smallest subfamily of brush-foots as it contains only the genus *Calinaga*, with species found in moist areas in the forests of South and East Asia. The taxonomy of this genus has not been fully resolved because of the species delimitations. However, recent efforts by researchers, including assembling comprehensive biogeographical and molecular data, have brought a more consistent approach to arranging the genus. Currently *Calinaga* comprises six species, *C. lhatso, C. buddha, C. brahma, C. aborica, C. formosana,* and *C. davidis*, with 30 valid subspecies. These butterflies are large and attractive, with pale grayish or cream coloration in the wings contrasting with a black body, some species showing bright orange or red scales in the thorax. Both sexes of most species and

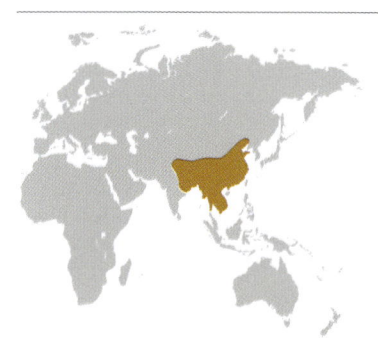

GENERA
Calinaga

DISTRIBUTION
Continental Southeast Asia

HABITATS
From tropical and temperate climates to near snowy peaks, at elevations of 300–10,000 ft (100–3,000 m).

SIZE
Large: ~1½ in (40 mm)

HOST PLANT FAMILIES
Moraceae

CONSERVATION
No species in the genus *Calinaga* are included in the IUCN Red List despite most being endemic with restricted ranges and therefore more vulnerable than others to any changes in the environment. *Calinaga buddha* is the only species in the genus that is currently protected under various schedules of the Wildlife Amendment Act

subspecies are very similar in appearance, although females can be slightly larger in size, paler, and rounder in the wings. Other traditional characters used to separate species in butterflies, such as male genitalia, are not well-studied or are not diagnostic. Most species are uncommon or extremely rare in the field and in collections worldwide, limiting studies on this genus. For example, *C. aborica* was first found in 1915, but researchers in India only found further specimens in the same locality 100 years later.

Although some species are localized, males can be found in dozens, puddling in wet areas in open forest and mountain hills; in contrast, females are seen less often and are unknown for some species. The immature stages are little studied in this genus; however, caterpillars of Asian *C. formosana* and *C. davidis* are elongated and green, with bright yellow spikes and prominent horns on the head. Pupae can also be green or brown.

in India. Citizen science initiatives and monitoring programs led by Indian researchers are key to recording species of conservation importance. No species in the subfamily have been listed by CITES

ABOVE | Orange-backed Freak *Calinaga brahma*. This unusual species inhabits primary forests near streams and rivers in India, Butan, and Myanmar in elevations up to 6,500 ft (2,000 m).

OPPOSITE | A female of the enigmatic Freak *Calinaga formosana*, a species found in Taiwan in elevations between 655 up to 6,500 ft (200–2,000 m).

LEAFWINGS, RAJAHS, PASHAS, AND EMPERORS

The Charaxinae has been a historically stable grouping. Adults are strong and fast flyers and have a broad, robust thorax and abdomen. The word "charaxes" means "sharp" or "pointed," and refers to the pointed tails on the rear hindwings of most species in the group. There are 19 genera grouping approximately 381 species and an impressive 836 subspecies that have been described. These butterflies seldom associate with flowers and instead feed on rotting fruit, putrefying fish, carrion, and dung. Many are magnificent, with multicolored, extremely beautiful, complex, or striking patterns on the upperside of the wings, including iridescence; the underside of the wing often provides camouflage, giving it the appearance of a dead leaf. Females are generally less iridescent. Some species are mimics of one another or of other butterflies. Eggs are usually globular, flattened at the top and bottom with only thin ridges. Larvae of most species are well-camouflaged to their surroundings, smooth, and lack spines or bright

ABOVE | The stunning Two-tailed Pasha *Charaxes jasius* is the only species of the large genus *Charaxes* found in Europe, and one of the largest butterflies on the continent.

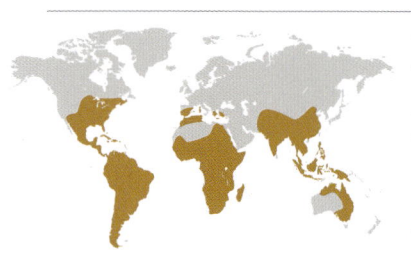

GENERA
Anaeomorpha, Palla, Agatasa, Prothoe, Charaxes (including *Polyura*), *Euxanthe, Archaeoprepona, Prepona* (including *Agrias,* and *Noreppa*), *Hypna, Coenophlebia, Zaretis, Siderone, Consul, Polygrapha, Phantos, Anaea, Fountainea, and Memphis*

DISTRIBUTION
Pantropical, Mediterranean, and Australia

HABITATS
Primary and secondary forests and gardens, up to 6,500 ft (2,000 m) elevation

SIZE
Medium to large: 1¼–2 in (30–50 mm)

HOST PLANT FAMILIES
Fabaceae, Lauraceae, Piperaceae, Sapindaceae, Euphorbiaceae, Mimosaceae, Annonaceae, Erythroxylaceae, Convolvulaceae, Monimiaceae, and Quiinaceae

colors. Some caterpillars will roll themselves inside leaves to hide when not feeding. Caterpillars have prominent, hard, chitinous protuberances on the head, referred to as "antlers."

Several tribes in this subfamily are no longer considered valid or to represent closely related

ABOVE | The Giant Charaxes *Charaxes castor* is widely distributed in African forests and savanna. Males are strong and fast flyers.

CONSERVATION

Bolivian *Agrias amydon boliviensis* is the only species in this subfamily listed in CITES Appendix III. Kenya's endemic *Charaxes nandina* is categorized as Vulnerable on the IUCN Red List. Some endemics of African countries include *Charaxes baileyi* and *C. nandina* (Kenya), *C. dowsetti* and *C. martini* (Malawi). The US Fish and Wildlife Service is revising the listing of North American *Anaea troglodyta floridalis* because of the recent decline in habitats, such as in the Everglades

groups. The genus *Anaeomorpha* is a peculiar lineage, including only a Neotropical species *Anaeomorpha splendida*, which has remained in its own tribe, Anaemorphini. Most Afrotropical and Asian species are currently placed in one genus, *Charaxes*, which is the largest, with 195 species and double the number of subspecies of the second largest. The popular Asian genus *Polyura* has been moved to become a subgenus of *Charaxes*, although this taxonomic arrangement is debated.

Two smaller tribes, Pallini and Prothoini, may continue to be recognized, with the transfer of *Agatasa* from the latter to the former. Butterflies in these tribes have an unusual morphological feature of serrated forewing edges. Individuals occasionally act aggressively toward one another, attacking each other's wings. This happens especially when males of the same or different species congregate at water sources or on mountaintops, behaviors referred to as "mud-puddling" and "hilltopping" respectively, which are prominent in this subfamily.

There are also two Neotropical tribes: the Anaeini whose species have a wing shape and often markings resembing a dead leaf; and the Preponini, which includes large brightly colored species in Neotropical forests, including the spectacular genus *Agrias* (popular among collectors), which has become a subgenus of *Prepona* following detailed morphological and molecular studies. However, these arrangements have not been fully implemented by researchers because of long familiarity with *Agrias* as a genus. Males are often

territorial and can be aggressive with other individuals. They usually live in the canopy or are seen in areas with more light within forests, such as along streams. Individuals will come down to ground level for food or nutrients and are attracted easily to baited traps.

OPPOSITE | The stunning *Prepona (Agrias) claudina* is found in tropical humid forests in the lowlands of South America. It is probably one of the most iconic and sought-after butterflies in the world.

BELOW | The genus *Agrias* contains some of the most beautiful butterflies in the Neotropics. Species in this genus have been reclassified recently into the genus *Prepona*; however, debate continues on this subject.

BELOW | The Blue Begum *Prothoe franck* inhabits forests in Southeast Asia and Australia. It can be found feeding on fermented fruits. Its underside (shown) contrasts with the dark brown and bright purple band in the upperside of the wings.

RINGLETS AND BROWNS

The satyrines are undoubtedly one of the most interesting groups among all Lepidoptera—contrary to how they might be perceived by the observer due to their generally dark or brown coloration. Satyrinae is the most species-rich subfamily of the Nymphalidae, with over 3,080 species and 3,700 subspecies—and probably many more yet to be described—arranged in 290 genera

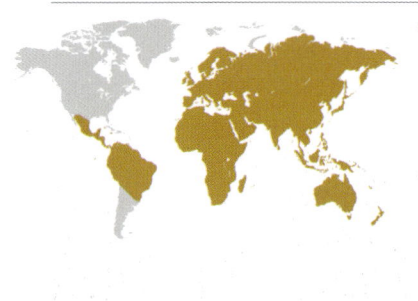

GENERA
Cithaerias, Dulcedo, Haetera, Pierella, and *Pseudohaetera* (Haeterini); *Aeropetes, Cyllogenes, Ducarmeia, Gnophodes, Haydonia, Manataria, Melanitis, Paralethe, Parantirrhoea,* and *Bletogona* (Melanitini); *Aphysoneura, Lethe, Neope, Ninguta, Ptychandra, Mandarinia, Chonala, Kirinia, Lasiommata, Lopinga, Orinoma, Pararge, Rhaphicera, Tatinga, Nosea, Callarge, Penthema, Zethera, Ethope, Neorina, Elymnias, Bicyclus, Brakefieldia, Culapa,* *Devyatkinia, Mycalesis, Mydosama, Lohora, Hallelesis, Heteropsis, Orsotriaena,* and *Telinga* (Elymniini)

DISTRIBUTION
All continents except Antarctica

HABITATS
Woodlands, forests, and savanna; African species are forest dwellers.

SIZE
Small to large: ½–1¾ in (14–45 mm)

and seven tribes. One of the characteristic features in this subfamily is the occurrence of at least one inflated basal vein in the forewings (although this can occur in Biblidinae too). They also have distinctive features in the wings, such as well-developed ocelli and, in many species, patterned, dark, cryptic undersides. Males of various species have androconial organs. Flying behavior across the subfamily varies widely, from species that fly only in sunny open areas to shady forest species flying low in the understory—there are even butterflies that only fly at dusk. Their eggs are spherical and laid individually or spaced out. Caterpillars can have horn-like structures on the head, and the pupa is suspended by a thread. Given the vast diversity of species and habitats for this subfamily, most characters that separate groups are non-universal, and there are always exceptions to the rule. However, immature stages and characters in male and female genitalia have proven to be useful for the separation of tribes. As with many other groups of butterflies, molecular characters provide new insights into classification.

The small Haeterini tribe includes five similar genera from the Neotropical region, encompassing 25 species. Among them is the charismatic Neotropical *Cithaerias*, whose species have almost

HOST PLANT FAMILIES
Poaceae; some specialized genera feed on Arecaceae (*Elymniopsis*).

CONSERVATION
The tribes here include range-restricted species: for example, most members of *Heteropsis* are endemic to Madagascar, and all species of *Lohora* are endemic to Sulawesi. Some Neotropical *Pierella* are inhabitants of pristine Amazonian forests currently being deforested. Despite this, no species are included in the IUCN Red List

OPPOSITE | The transparent butterfly *Pseudohaetera hypaesia* is found in mid-elevation forests in Andean countries in South America. Males display territorial behavior.

TOP | The fruit-feeding Dusky Evening Brown *Gnophodes chelys* is a common butterfly from Africa, active only at dawn and dusk.

ABOVE | The Dark-branded Bushbrown *Mycalesis mineus* has been used by teams of researchers as a model organism to investigate the evolution of phenotypic plasticity in butterflies.

ABOVE LEFT | The Common Evening Brown *Melanitis leda* is a widespread species found in different habitats from forests to open grasslands in Africa, Asia, and Australia. It displays multifaceted polymorphism in wing coloration between dry and wet seasons.

ABOVE | The Common Palmfly *Elymnias hypermnestra* is found in South Asia and displays a strong marked dimorphism. Males are blackish-brown, while females are orange, mimicking other species.

LEFT | The Wall *Lasiommata megera* gets its name because of its habit of basking on walls, rocks, and stony places. This species can be found in Europe, North Africa, and East Asia.

transparent wings with shades of pink and purple. Butterflies in this group are inhabitants of pristine wet forests and glide near fallen leaves covering the soil.

The tribe Melanitini is mostly distributed in tropical Africa and Asia, although the genus *Manataria* is the only one in the Neotropics; in contrast, the genus *Melanitis* is widespread across the Eastern Hemisphere, but not the Americas. There are ten genera and 33 species known. Some species can be pests of crops: *Melanitis leda ismene* is known as the rice butterfly.

The Elymniini tribe is distributed around the world except for the Americas. It includes over 600 species arranged in 32 genera, and incorporates large former tribes, such as the African Mycalesini.

Ranging from monotypic genera, such as Asian *Mandarinia regalis* and China's beautiful endemic *Nosea hainanensis*, to the most diverse and species-rich, such as African *Bicyclus*, its hundred-plus species are favorites for phylogenetic and morphological studies. Interestingly, species can be arranged according to the position and detail of their androconia. Most species are brown although diversity is so vast that there are also white butterflies that do not look like a typical member of Satyrinae, for instance the African *Aphysoneura pygmentaria*. This tribe includes large butterflies with the crepuscular habit of flying only at dawn and at dusk, such as the Asian genus *Elymnias*.

GRAYLINGS, MEADOW BROWNS, RINGLETS, ARGUSES, AND BANDED SATYRIDS

This cosmopolitan group includes medium-sized butterflies occurring in basically any habitat on Earth, but not at the poles. This tribe is more diverse than any other, with more than 200 genera, over 2,000 species, and another 2,000 subspecies that have been described, with many more yet to be named. Discoveries are occurring as this book is being written, resulting in ongoing changes in the taxonomy of this subfamily. Morphological differences in adults of several brown-looking species are small at first inspection, such as in the complex, recently studied Neotropical subtribe Euptychiina; with the added issue of some species having seasonal variation,

LEFT | The Marbled White *Melanargia galathea*, a charismatic and widespread species in the Palearctic region, has garnered significant attention for its varied wing patterns. This has resulted in the naming of hundreds of variations, many of which are now considered synonyms.

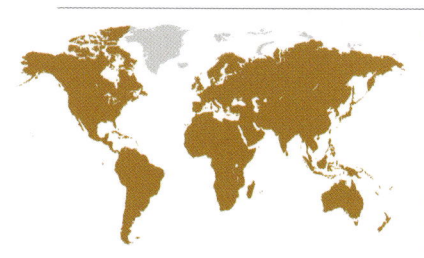

GENERA
Genera include: *Altiapa, Tisiphone, Hyponephele, Maniola, Melanargia, Corades, Lasiophila, Lymanopoda, Mygona, Neopedaliodes, Oxeoschistus, Steroma, Steromapedaliodes, Thiemeia, Gyrocheilus, Arethusana, Aulocera, Berberia, Boeberia, Brintesia, Cassionympha, Chazara, Coenyra, Coenyropsis, Davidina, Hipparchia, Melampias, Neocoenyra, Oeneis, Paralasa, Paroeneis, Pseudochazara,* *Pseudonympha, Paternympha, Satyrus, Strabena, Ypthima, Coelites, Acrophtalmia, Acropolis, Coenonympha, Lyela, Triphysa, Sinonympha, Calisto, Tarsocera, Dingana, Callerebia, Argestina,* and *Erebia*. A detailed list of genera in the subtribe Euptychiina is found in Zacca et al.

DISTRIBUTION
All continents except Antarctica

HABITATS
Primary and secondary habitats, including woodland, tropical forest, savanna, paths,

delimiting species is very difficult. Recent molecular studies and collaborative initiatives of integrative taxonomy (integrating character morphology and DNA) have doubled the numbers of known genera and species over the last decade. For a curated checklist of the subtribe Euptychiina, visit the Euptychiina Checklist web page (see Resources).

Most species are fruit feeders, with many attracted to fermented fruits, carrion, and dung, although species in African *Ypthima* visit flowers. Palps on some species have long, hair-like scales on the edge. Despite the dull coloration of most

and parks; from sea level to high elevations above 10,000 ft (3,000 m) in the mountains of South America

SIZE
Small to medium: ⅔–1¼ in (15–30mm)

HOST PLANT FAMILIES
Poaceae; some genera feed on Marantaceae, Arecaceae, Cyperaceae, and Selaginellaceae

CONSERVATION
The African species *Dingana fraternal* and *Stygionympha dicksoni* and European *Pseudochazara cingovskii* and *Coenonympha phryne* are categorized as Critically Endangered on the IUCN Red List. Brazilian endemic *Euptychia boulleti* is listed as Critically Endangered in the Brazilian Red Book of Endangered species. *Neonympha mitchellii* is listed as Endangered in the Eastern USA

TOP | Among the predominantly brown-colored euptychiines, *Caeruleuptychia urania* gets its name for its blue coloration.

ABOVE | The Swordgrass Brown *Tisiphone abeona* is endemic to Southeastern Australia. In some high-elevation populations there is only one generation annually.

butterflies in this tribe they are of special conservation interest because of the limited ranges, mostly in high mountains, many species have. There are hundreds of endemic genera in this tribe, such as the Neotropical *Lymanopoda*, *Idioneurula*, *Manerebia*, *Carminda*; China's endemic *Sinonympha*; and New Caledonia's *Austroypthima* to name a few. Other genera are more widespread, such as the almost global *Ypthima* and Palearctic *Erebia*, which have over 100 species each, and are difficult to tell apart in many cases. The males and females of most species are alike, with small differences in size, wing shape, and coloration, but they are distinguishable because of the presence of androconial scales in the wings of males; these structures can also be used to separate groups as well as to understand evolutionary traits, as in the Palearctic genus *Pseudochazara*.

Although the appearance of members of this tribe is usually unremarkable, there are species with shades of white, purple, and orange, and some have metallic coloration too, for instance Colombia's endemic *Lymanopoda caeruleata*.

Eggs are round and often laid solitarily or, rarely, in pairs. Caterpillars have a bifid tail and can be found feeding in groups on monocotyledonous plants, such as bamboos and grasses. In species of Palearctic areas with marked seasons, the winter months are passed at the caterpillar stage; an example is the European Small Heath *Coenonympha pamphilus*.

OPPOSITE | A pair of Common Four Ring *Ypthima huebneri*, a small and often ignored butterfly because of the drab coloration on the upperside of its wings.

LEFT | The Meadow Brown *Maniola jurtina* is a widespread butterfly in the Palearctic region and North Africa. This species has become a model study for geneticists because of the variations in the spotting pattern on its wings.

BELOW LEFT | The rare *Lymanopoda nevada*, an endemic species of Colombia, found only on one mountain in bamboo forests above 9,800 ft (3,000 m).

MORPHOS, ANTIRRHEA, AND CAEROIS

BELOW | The iridescent female *Morpho cypris* is an incredibly elusive butterfly, making it a rare and challenging subject to spot and photograph in the wild.

The morphos are among the best-known and most spectacular butterflies in the world. This genus comprises 33 species and as many subspecies, which are restricted to the Neotropical region. The genus *Morpho* was considered so unique that it was placed in its own subfamily many years ago. However, more detailed studies—including of immature stages— demonstrated that two other and smaller Neotropical genera, *Antirrhea* (currently

GENERA
Morpho, Antirrhea, and *Caerois*

DISTRIBUTION
Neotropical South and Central America

HABITATS
From sea level up to 7,000 ft (2,200 m) in forested habitats with clearings and decaying wood, and near streams

SIZE
Medium to very large: 1¾–4¼ in (45–105 mm)

HOST PLANT FAMILIES
Fabaceae, Poaceae, Lauraceae, and Euphorbiaceae (*Morpho*); monocotyledons including palms and grasses in families Musaceae and Arecaceae

CONSERVATION
Two Brazilian subspecies *Morpho epistrophus nikolajewna* and *M. menelaus eberti* are listed as Critically Endangered in the Red Book of Endangered Fauna of Brazil. Dramatic reduction, conversion, and

13 species) and *Caerois* (two species), are part of the same group, supported by molecular studies that have now placed this group inside the large Satyrinae subfamily. Most species in the genus *Morpho* have iridescent color, often blue, but sometimes orange or whites; some reflect almost all spectrums of light with a silvery or metallic effect, such as the Andean species *Morpho sulkowskyi*. In contrast to the bright colors on the upperside, most species have dark patterns on the underside that provide camouflage, and marks resembling eyespots or ocelli. Morphini includes some of the largest species found in the Americas.

Popular species include the Blue Morpho *Morpho helenor*; however, not all species are blue. The Sunset Morpho *M. hecuba*, one of the largest species, has dark orange-and-white

RIGHT | A female of White Morpho *Morpho polyphemus*, a species found from Mexico to Costa Rica. Because of their large size and beauty, this species is bred in live butterfly houses to engage the public.

BELOW RIGHT | The Blue Morpho *Morpho helenor* is a large and charismatic species found throughout the Neotropical region. There are over 30 recognized subspecies, some of which are widespread, while others are more locally restricted.

loss of forested habitats are listed as the most important causes of threat of extinction in Brazilian forests. The subspecies *M. godartii lachaumei* is listed on CITES Appendix III

flashes on its wings. The Central American White Morpho *M. polyphemus* is, as the name suggests, uniformly white, including its underside. The coloration of these butterflies is not due to pigmentation but arises instead because the tiny scales in the wings reflect and refract light. Blue Morphos are popular study subjects for light reflection as well as for the development of products using iridescence, such as paint.

The three genera of Morphini share an interesting feeding ecology. As adults, they have a diet of rotting or fermented fruit, and in butterfly houses adults can be fed with bananas. *Antirrhea* and *Caerois* feed on fruits in palm trees, but only those old enough to have developed a fungal covering, congregating on occasions when there is an abundance of such fruit. Unusually, none of these butterflies appears to feed on nectar.

Morphos often show preferences for specific forest levels, or strata, during flight, with some species, such as the spectacular *M. cypris*, flying well above the canopy. Unlike most *Morpho*, species in *Antirrhea* are more active at dusk and do not fly in direct sunlight, instead keeping close to the forest floor. *Antirrhea* are moderately large and usually brown with purple or orange marks in the wings. They have particularly marked brush-like androconial organs in the overlapping sections of the wings and marked scent organs on the genitalia. Species in the genus *Caerois* have unusual wing shape and orange-and-black color.

Morphos have been well-studied and are popular in butterfly houses, where several species may be reared. As a result, their immature stages and lifecycle are well-known, however, little is known about species in *Antirrhea* and *Caerois*. *Morpho* females lay single eggs on the underside of leaves. The eggs of all three genera are unusual in being hemispherical. Caterpillars are colorful with tones of bright red and yellow, hairy, and with a

OPPOSITE | Although most iconic *Morpho* butterflies have attractive colored patterns on the upperside of their wings, the ventral side of the wings have mildly colored scales and eyespots assisting with camouflage.

LEFT | A male of *Antirrhea philoctetes*, a species that inhabits pristine forests in the Neotropical region. Butterflies in this genus are uncommon in the wild and in collections.

LEFT | *Caerois chorinaeus* is an uncommon species found in the Amazon. Males of this species display intricate and specialized scent organs on the wings.

brushy gland that can emit a pungent odor when an individual is threatened. The pupa in *Morpho* is large and green, and it is usually suspended from a narrow branch, appearing like a fruit, being either ovoid or angled.

OWLS

The Brassolini were previously considered a subfamily but, as with the closely related Morphini, are now classified as a tribe within the Satyrinae. They include medium to very large, usually brownish butterflies with subtle tones of blue, purple, orange, and white colors on the upperside of their wings. The undersides include large ocelli or spots, hence species in *Caligo* commonly being called owls, a name that might also be associated with them being more active at dawn and dusk (which is very unusual among butterflies), when it is difficult to see. All species occur in the Neotropical region only. This tribe currently comprises 17 genera, over 100 species, and more than 200 subspecies. However, molecular studies suggest that it could be rationalized further.

GENERA
Bia, Blepolenis, Brassolis, Caligo, Caligopsis, Catoblepia, Dasyophthalma, Dynastor, Eryphanis, Mielkella, Opoptera, Opsiphanes, Orobrassolis, Penetes, Selenophanes, Aponarope, and *Narope*

DISTRIBUTION
Neotropical Central and South America at 650–9,000 ft (200–2,800 m) elevation

HABITATS
Lowland forests, foothills, and Amazonian and Andean forests; palm, sugarcane, and banana plantations

SIZE
Medium to large: 1–3 in (25–75 mm)

The genera *Caligo* (21 species), *Narope* (17), and *Opsiphanes* (13) are considered the most diverse; several other currently recognized genera contain only a single species. The genera *Caligo* and *Dynastor* contain the largest-sized species in the Americas.

Being so large and relatively colorful, butterflies in this tribe have attracted the attention of butterfly enthusiasts; *Caligo telemonius memnon* is often bred and sought after for butterfly houses, and its life history has been well-studied. However, most other genera in Brassolini remain little known, with many species lacking a documented life history and resolved taxonomy, with new species being discovered in recent years. The Amazonian leaf-shaped genus *Bia* is now considered to

---

HOST PLANT FAMILIES

Cannaceae, Marantaceae, Monimiaceae, Rubiaceae, Zingiberaceae (*Caligo*), Araceae (*Opsiphanes*), Cycadaceae (*Opsiphanes*), Malvaceae (*Brassolis*), Heliconiaceae (*Caligo, Opsiphanes*), Arecaceae (*Caligo, Opsiphanes, Brassolis*), Musaceae (*Caligo, Opsiphanes, Brassolis*), and Poaceae (*Caligo, Opsiphanes, Brassolis, Eryphanis, Narope*)

CONSERVATION

Brazilian endemic species *Orobrassolis ornamentalis*, *Dasyophthalma geraensis*, and *Dasyophthalma rusina delanira* are listed as Critically Endangered in the Red Book of Endangered Fauna of Brazil, while *Dasyophthalma vertebralis* is listed as possibly Extinct. There are no species in this tribe listed in CITES

OPPOSITE | This owl butterfly *Eryphanis zolvizora* is a rare Andean endemic historically overlooked by researchers and naturalists as a single species until recently.

ABOVE | The Owl *Caligo atreus*, a large Neotropical butterfly named because of the large eyespots on the back of its hindwings. They fly at dusk and dawn.

comprise six species following a recent revision, from two originally recognized. Historically, it has been placed in Morphinae or in its own tribe, but it is now considered an older lineage of Brassolini. Subspecies of the Andean *Eryphanis zolvizora* increased from three to eight in a recent study. Owl butterflies have well-developed, specialized androconial scales on the wings.

Although most are forest dependent—including endemic species such as *Orobrassolis ornamentalis* (Brazil)—some species can often be found in banana plantations and emergent secondary forests with abundant palm trees. Several species are important pests at their caterpillar stage—for example, *Opsiphanes cassina* (oil palm), species in the genus *Brassolis* (coconut), and *Caligo* (banana and sugarcane). The life histories of some brassolids—especially *Caligo telemonius memnon*—are well-known. Eggs of butterflies in this tribe are of varied shape and size—rounded, smooth, or more ridged—and can be laid singly or in clusters. Caterpillars are mostly gregarious, including those of *Caligo* species, but caterpillars of *Opoptera syme* are not. They have prominent heads and horns and a swallow-like forked tail, which can be quite long. Coloration varies across the different instars. Pupae in *Narope cyllene* are usually dark amber in color with a rough surface, but in *Brassolis isthmia* are bright yellow and smooth, and in *Caligo atreus* are bright red and resemble a sarcophagus.

FAWNS, SATURNS, JUNGLE QUEENS, SILKY OWLS, CALIPHS, DUFFERS, AND LEAFS

The Amathusiini is a diverse group of rather big, often colorful butterflies that look like dead leaves or have soft brown camouflage patterns on the underside of the wings. They occur only in Asia, with the greatest concentration of species in the tropical region. One radiation of this group can be thought of as the Asian counterparts of the Morphini (morphos) and Brassolini (owls), with some genera having similar morphology, and frugivorous, crepuscular ecology. There are approximately 115 species and 383 subspecies in this group as currently recognized, across 13 genera. Not long ago, genera *Hyantis* and *Morphopsis* were included in the tribe, but are now considered either as part of the Melanitini or possibly in their own group, the Hyantini. Finally, the genus *Xanthotaenia* was also once placed in the Amathusiini, but is now considered part of the Zetherini. A definition of this group has been controversial, although the present arrangements follow very recent molecular studies.

The most ancient genus in this group is the Asian *Stichophthalma* or jungle queens, which comprise 14 species. According to both molecular and morphological studies, these are the most distantly related genus to others in the tribe. They are particularly pleasing in appearance, being large butterflies with white, light orange, dark brown, or purple colors. Several species have black diamond shapes arranged around the edges of the wings and

RIGHT | The patterns on the underside and upperside of the wings in *Stichophthalma* vary greatly among species and subspecies.

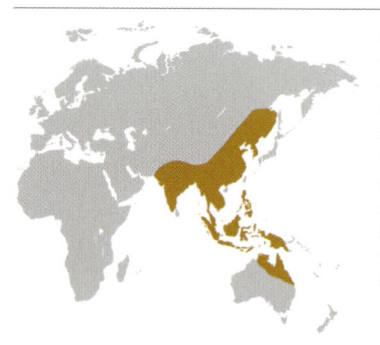

GENERA
Stichophthalma, Faunis, Allaemona, Aemona, Melanocyma, Enispe, Discophora, Amathusia, Zeuxidia, Amathuxidia, Thaumantis, Thauria, and *Taenaris*

DISTRIBUTION
Asia, from India and China through to Southeast Asia and islands to Borneo, northern Australia, and New Guinea; tropical areas, usually at lower elevations

HABITATS
Mostly lowland forests; palm and banana plantations

SIZE
Medium to large: 1–2⅓ in (28–60 mm)

HOST PLANT FAMILIES
Arecaceae, Asparagaceae, Costaceae, Cycadaceae, Hypoxidaceae, Musaceae, Pandanaceae, Poaceae, Smilacaceae, and Zingiberaceae

CONSERVATION
Overall, IUCN Red List assessments have been conducted mostly on European species, omitting tropical species in Amathusiini and other groups. However, this does not mean there are no threatened species; some unique species are vulnerable, such as Sulawesi's endemic butterfly *Faunis menado*. No species are listed on the CITES checklist

TOP | Despite the restricted distribution of Jungle Queen *Stichophthalma mathilda* in Southeast Asia, the wing patterns and genitalia are variable, which makes naming and classification challenging. This appears to be an undescribed subspecies.

ABOVE LEFT | A female *Zeuxidia amethystus*. This species is found in primary forests near streams. Adults are found flying in early morning or late afternoon.

ABOVE RIGHT | A male *Faunis canens*, a species found in pristine forests in low elevations in some countries in Southeast Asia. Adults fly close to the ground.

eyes on the underside, while others have more simple wing bands. Genera *Zeuxidia* and *Amathuxidia* also include large and spectacular species. In contrast, the fawns *Faunis* and *Allaemona* are mostly soft brown butterflies that often lack distinctive marks, colors, or contrasts in their scale coloration. Apparently closely related to the fawns, but very different morphologically, is the genus *Taenaris*, which includes the so-called silky owl butterflies. These are very striking in appearance, with white undersides and huge orange-ringed ocelli on the hindwings, which in some species are perhaps the most prominent "eyes" of any butterfly, contrasting strongly with the pale base color of the wings.

While most genera in Amathusiini rely upon camouflage to evade predators, these species are showy, with the aim of giving off a warning of toxicity to predators. Their larvae feed on toxic host plants that contain cyanide-based compounds. Some species form mimicry rings, where different species occurring in the same geography mimic each other's similar striking patterns to collectively

warn predators of their non-palatability. The caterpillar of the Palm King *Amathusia phidippus* lays its eggs in palms and is regarded as an agricultural pest because it feeds on coconut palms. Males in *Zeuxidia* can be seen during daylight hours within dense forest, feeding on fallen fruit, features paralleled by the forest habitats, color patterns, and frugivory of *Morpho* in South America.

The sole species of *Amathuxidia* is called the Koh-i-noor, named for the famous diamond. This is also a leaf-like butterfly with overall similar patterning to *Zeuxidia*, but it is more widely distributed, including in India.

Widespread species in the complex *Amathusia phiddipus* are difficult to identify because of the similarity in wing patterns. Taxonomists often use secondary sexual characters found in the males to assist correct identification.

The charismatic *Melanocyma faunula* is a species found in the Malay Peninsula. Sexes are alike and large in size, but males have specialized scales on the upperside of the wings.

Species in the Asian genus *Discophora* have drab coloration and camouflaged patterns on the underside of the wings; however, on the upperside, males and females look distinctive.

GLOSSARY

Accessory glands: Internal reproductive organs present in both sexes that play a vital role in reproductive strategy.

Androconia: Specialized scales found in male butterflies involved in dissemination of scents or pheromones. The term was introduced by Samuel Scudder.

Androconial scales: See *androconia*.

Antennae: Paired and segmented appendages on the head, involved in odor-sensing and navigation. Butterfly antennae are typically clubbed.

Apices: Plural of apex, the top or highest part of the wing.

Aposematic: See *aposematism*.

Aposematism: A phenomenon where unpalatable organisms use bright warning colors to signal their distastefulness to predators.

Bifid: Forked, or divided into two. The word derives from bi (two) and -fid (stem of findere, meaning "to separate").

Bipectinate: Feather-like antennae resulting from having multiple processes on both sides of the antenna.

Camouflage: Defense mechanism for hiding from predators, typically by blending into the surroundings.

Chorion: Ribbed or ridged outer layer of the egg.

Chrysalis: The butterfly pupa; the stage between larva and adult.

CITES: The Convention on International Trade in Endangered Species of Wild Fauna and Flora.

Clade: A group of animals composed of a common ancestor and all of its descendants.

Cladistic: A system of categorizing based on inferred evolutionary relationships and branching patterns without a time scale.

Complete metamorphosis: The process of passing through the four life-stages: egg, larva, pupa, and adult.

Costal: Anterior-most part of the forewing, which starts at the wing base and terminates at the apex.

Cubitus: Wing vein forming the posterior margin of the discal cell, divided distally.

Cupola organ: Larval glands that play an important role in maintaining the ant association mainly in Lycaenidae and Riodinidae.

Cuticle: Chitinized or hardened outer layer of the body, serving as a physical barrier.

Eclose: Emergence of adult from pupa/chrysalis, or larva from an egg.

Ectotherm/y: Animal whose body temperature depends on an external heat source, also known as "cold blooded."

EDGE program: The Evolutionarily Distinct and Globally Endangered (EDGE) species program; a research and conservation initiative.

Endemic: Confined to a narrow range, often only found in a single geographical region, e.g., islands.

Eversible: Can be everted, i.e., turned inside out.

Filiform: Thread-like antennae composed of evenly sized segments.

Fimbriae: Fine threads or projections resembling a fringe in the apex of wings.

Form (as used in this book): Individuals that differ from other individuals within a taxon.

Frenulo-retinacular: A type of wing-coupling mechanism common in moths where the wings are hooked with a frenulum.

Frenulum: A bristle-like structure on the hindwing that serves as a "hook" to connect to the forewing.

Frugivory: Consumption of fruits; butterflies that are primarily fruit-eating.

Guild: Grouping based on resource requirements, such as feeding preference.

Hair pencil: Brush-like scent organs. They can be found in various parts of the male body, and are involved in dispersing scent or pheromones.

Hemolymph: Extracellular fluid in invertebrates that functions like blood in vertebrates.

Hilltopping: Behavior performed by males congregating on topographical prominences (e.g., a hill summit) in order to find females for mating.

Instar: A development stage in arthropods occurring between molts until sexual maturity.

IUCN: The International Union for Conservation of Nature.

Lamina: Thin membranous layer that forms the main structure of the wings.

Mimetic: Imitating in appearance another organism by mimicry (see below).

Mimicry: A survival strategy where insects resemble other species, in some cases ones that are unpalatable to predators.

Monophyly: Referring to a group of organisms that share a unique common ancestor.

Monotypic: A taxonomic group that only has one taxon, i.e. only one representative in a genus, only one species, or, in a family, only one genus.

Mud-puddling: A behavior displayed in insects, more commonly in butterflies, where they congregate on moist ground to obtain nutrients.

(Multi-)voltine: Possession of multiple broods or generations annually.

Myrmecophily: A mutualistic association between ants and other organisms.

Nanostructures: Materials or objects that have at least one dimension in the nanometer scale.

Ocelli: Eyespots found on the wings of butterflies that vary in shape, number, color, and position.

Ommatidia: Individual optical units that make up the compound eyes of insects.

Osmeterium: A glandular organ found in the first thoracic segment in swallowtail caterpillars; it emits odors that deter predators.

Palpi: A pair of segmented appendages usually covered in scales and sensors, located in the head near the mouthparts of butterflies.

Phytophagy: Feeding on plant material.

Polymorphism: The occurrence of two or more forms or phenotypes within a species.

Pouch: A particular structure or specialized adaptation, such as an androconial pouch to release pheromones.

Pretarsal: The terminal segment of the tarsus and structures attached to it in an insect's leg.

Proboscis: A straw-like, elongated feeding structure found on the heads of Lepidoptera, used to obtain nectar and other substances from flowers and plants.

Puddling: See *mud-puddling*.

Pupa: The life-stage between the larvae and the adult stage where the insect undergoes metamorphosis.

Retinaculum: Structure involved in wing coupling.

Scoli: A long, horn-like structure with tiny, scattered projections attached to the head of caterpillars.

Setae: Bristle or hair-like structures that aid in defense against predators; also used for insulation and heat exchange.

Sexual dimorphism: Morphological differences between males and females of the same species or subspecies, such as wing patterns and size.

Spiracles: Part of the respiratory system, spiracles are small openings on butterflies' abdomens that are used to exchange gases and respire.

Spur: Socketed structure found on tibia, sometimes moveable.

Superfamily: A taxonomic ranking below order, which includes groups of families considered to be closely related by a common ancestor. The standard ending of a superfamily's name is "oidea."

Sympatry: Occurrence of two or more species living in the same geographic area.

Thorax: The middle part of a butterfly's body, between the head and the abdomen, containing all the muscles required for moving, flying, or walking.

Three-dimensional structures: Solid structures containing three different dimensions such as length, width, and height.

Tibia: A segment in the leg of an insect located between the femur and the tarsus.

Tubercles: Small projections or outgrowths found in the bodies of caterpillars.

Xeric: An environment with very low moisture levels. Its name comes from the Greek "xeros" meaning dry.

RECOMMENDED RESOURCES

BOOKS

Asher, J., Warren, M., Fox, R., Harding, P., Jeffcoate, G. & Jeffcoate, S. (Eds.). *The Millennium Atlas of Butterflies in Britain and Ireland*. Oxford University Press, 2001.

Bauer, E. & Frankenbach, T. *Butterflies of the World*, various volumes. Goecke & Evers, 1998–2022.

Benyamini, D. & John, E. *Butterflies of the Levant and Neighbouring Areas*, volumes 1–4. 4D Microrobotics, 2016–24.

Bozano, G. C. (Ed.). *Guide to the Butterflies of the Palearctic Region*, parts 1–5. Omnes Artes, 2003–18.

Braby, M. F. *Butterflies of Australia: Their Identification, Biology and Distribution*, volumes 1 & 2. CSIRO Publishing, 2000.

Corbet, A. S., Pendlebury, H. M., & Eliot, J. N. *The Butterflies of the Malay Peninsula*. Malaysian Nature Society, 2020.

DeVries, P. J. *The Butterflies of Costa Rica and Their Natural History*, volumes 1 & 2. Princeton University Press, 1987 & 1997.

Freitas, A. V. L., et al. *Livro Vermelho da Fauna Brasileira Ameaçada de Extinção*, volume VII – Invertebrados (Red Book of Endangered Brazilian Fauna. IUCN, 2018.

Hall, J. P. W. *A Monograph of the Nymphidiina (Lepidoptera: Riodinidae: Nymphidiini): Phylogeny, Taxonomy, Biology, and Biogeography*. Entomological Society of Washington, 2018.

Hall, J. P. W. *A Phylogenetic Revision of the Napaeina (Lepidoptera: Riodinidae: Mesosemiini)*. Entomological Society of Washington, 2005.

Huertas, B., Le Crom, J. F. & Correa-Carmona, Y. *Endemic Butterflies of Colombia*. Natural History Museum London & ProColombia, 2022.

Klimaitis, J F., Núñez Bustos, E. O., Klimaitis, C. L. & Gueller, R. M. *Mariposas de Argentina: Guía de Identificación (Butterflies of Argentina: Identification Guide)*. Vazquez Mazzini Editores, 2018.

Lamas, G. (Ed.). *Atlas of Neotropical Lepidoptera*: Part 4A. Hesperioidea – Papilionoidea. Scientific Pub, 2004.

Larsen, T. B. *Butterflies of West Africa*, volumes 1 & 2. Apollo Books, 2005.

Le Crom, J. F., Salazar Escobar, J. & Constantino Chuaire, L. M. *Mariposas de Colombia*, volumes 1–3. Carlec, 2002, 2004 & 2023.

Monastyrskii, A. L. *Butterflies of Vietnam*, volumes 1–5. Dolphin Media Company, 2005–24.

Nakae, M., Nishiyama, Y. & Cotton, A. *Papilionidae of the World*. Roppon Ashi, 2021.

Neild, A. F. E. The *Butterflies of Venezuela*, volumes 1 & 2 Meridian, 1996 & 2008.

Scott, J. A. *The Butterflies of North America: A Natural History and Field Guide*. Stanford University Press, 1986.

Tolman, T. & Herrington, R. *Collins Butterfly Guide: The Most Complete Guide to the Butterflies of Britain and Europe*. Collins, 2009.

Tshikolovets, V. V. et al. *The Butterflies of Palaearctic Asia Series*, volumes 1–10. Tshikolovets Publications, 2016–20.

Tshikolovets, V. V. *The Genus* Calinaga Moore, *[1858]*. Tshikolovets Publications, 2020.

Tsukada, E. (Ed.). *Butterflies of Southeast Asian Islands – Illustrated Guide Book*, volumes 1–3. Plapac, 1982–85.

Tyler, H. A., Brown, K. S. & Wilson, K. H. *Swallowtail Butterflies of the Americas: A Study in Biological Dynamics, Ecological Diversity, Biosystematics, and Conservation*. Scientific Pub, 1994.

Willmott, K. R. *The Genus Adelpha: Its Systematics, Biology, and Biogeography*. Scientific Pub, 2001.

WEBSITES

Afrotropical Butterflies and Skippers: A Digital Encyclopedia: lepsocafrica.org

Butterflies of America: www.butterfliesofamerica.com

Butterflies and Moths of North America: www.butterfliesandmoths.org

Butterfly Catalogs (Neotropics): www.butterflycatalogs.com

Butterfly Conservation UK: butterfly-conservation.org

Checklist of Butterflies in Indochina: yutaka.it-n.jp

Euptychiina Checklist (Neotropics) (Zacca et al.): labbor.ib.unicamp.br

Butterflies of India: www.ifoundbutterflies.org/papilio-demoleus

IUCN Red List of Threatened Species, Version 2024-2: www.iucnredlist.org

Learn About Butterflies: learnbutterflies.com

Lepidoptera and Some Other Life Forms: https://ftp.funet.fi/index/Tree_of_life/intro.html

The Lepidopterists' Society: www.lepsoc.org

Natural History Museum (UK) Data Portal: data.nhm.ac.uk/search

Nymphalidae Phylogeny: www.nymphalidae.net/Nymphalidae/Phylogeny/Phylogeny.htm

Petit's Ecuador Butterflies: www.sangay.eu/en and www.cotacachi.eu/en

Tree of Life Web Project: tolweb.org/tree/phylogeny.html

INDEX

Mimic 200
Mimic Flat 89
mimicry 32–33, 71, 80, 98, 185, 230
Ministrymon azia 159
"The Mirror" 104, 105
models, for studies 115, 117, 183, 219
Monarch butterflies 8, 19, 21, 25, 164, 173
Morphini 58, 220–223, 228
Morpho 220, 221, 223
Morpho cypris 220
Morpho hecuba 221–222
Morpho helenor 221, 222
Morpho menelaus 42
Morpho peleides 60
Morpho polyphemus 221, 222
Moschoneura sp. 119
moth-like butterflies 11, 31, 82–85
Mother-of-Pearl 201, 202, 203
moths, butterflies *vs* 11, 17–18
Müllerian mimicry 32–33, 185
museomics 46
museum specimens 40, 46, 47, 49, 52, 59, 125
Mycalesis mineus 213
myrmecophily *see* ant(s), caterpillar relationship
Myscelus 99
Mysoria 88

N

Nabokov, Vladimir 51, 53
Nagoya Protocol on Access and Benefit Sharing 48–49
Narope anartes 226, 227
Natewa Swallowtail 31
Natural History Museum, London 47, 57
natural selection 8, 40
naturalists 8, 30, 41, 44, 45, 49
Nearctic region 28, 29
nectary organ 151
Nemeobiinae 133, 136–137
Nemeobiini 136
Neotropical region 13, 26, 28, 31, 38, 41, 46, 49
Neptini 176
Neptis 174, 176
Neptis pseudovikasi 176
Nessaea aglaura 193
Netrocorynini 96–97
Nettle-tree Butterfly 167, 168, 169
new species: discovery 52–53, 59, 61
  naming 50–51, 53
Ninja 186–187
nocturnal butterflies 84

nomenclature 50–51, 60
Norse Grayling 22
Northern Chequered Skipper 86, 87
Nymphalidae 6, 11, 15, 18, 58, 163–231
  camouflage 33
  eggs, caterpillars, pupae 20, 21, 23
Nymphalinae 200–205
Nymphalini 202
Nymphidiini 140

O

Oakleaf 33, 34
Ochlodes sylvanus 110
Odontoptilum angulata 97
Oeneis 20
Oeneis norna 22
Oileidini 94–95
Old World Swallowtail 7
ommatidia 18
online initiatives, studying butterflies 49
Opoptera aorsa 226, 227
Opsiphanes cassina 226
Orange Awlet 90
Orange-backed Freak 207
Orange Hairstreak 34
Orange-tip 130
Orcus Checkered-Skipper 101
Oriental Map butterfly 199
Ornipholidotos 152, 153
Ornithoptera 28, 30, 76, 77
Ornithoptera alexandrae 30, 45, 48, 58, 77
Ornithoptera croesus 77
osmeterium 34, 64, 67
Ourocnemis renaldus 135
overwintering 20, 22
Owl (*Caligo atreus*) 225, 226
owls (*Caligo*) 35, 225
Oxylini 156
Oxynetrini 98–99

P

Paches loxus 101
Painted Jezebel 127
Painted Lady 28, 164, 202, 203
paintings of butterflies 42, 43, 54–56
Pale Green Awlet 90, 91
Palearctic region 28, 29
Pallini 210
Palm King 230, 231
Papilio 11, 43, 78–79, 80
Papilio antimachus 80
Papilio dardanus 8, 78, 80
Papilio demodocus 80
Papilio demoleus 80
"*Papilio ecclipsis*" (hoax butterfly) 122
Papilio flavomarginatus 41
Papilio krishna 65

Papilio lowi 79
Papilio machaon 7, 79
Papilio memnon 64, 79
Papilio morondavana 29
Papilio natewa 31
Papilio zelicaon 80
Papilionidae 10, 11, 17, 20, 44, 50, 58, 62–81
Papilioninae 20, 50, 63, 72–80
Papilionini 50, 72, 78–80
Papilionoidea 10–11, 50, 58, 61, 89
Pardopsis punctatissima 181
Pareronia 131
Parides 28, 76, 77
Parides ascanius 31, 77
Parnassiinae 17, 63, 68–71
Parnassius 29, 69, 70
Parnassius apollo 69, 71
Parthenini 177
Parthenos 177
Parthenos sylvia 177
Passovini 98–99
Patia cordillera 118
patrolling 24, 102
Peacock 164
Peacock Royal 157
Pentilini 152–153
Perisama 197
Perisama humboldtii 197
Peru, species diversity 26
Phanus marshalli 94
Phellinodes spp. 82, 83
Phengaris arion 161
pheromones 15, 24
Philaethria dido 183
Phocidini 94–95
photography, butterfly 27, 57
photonic crystals 72
photoreceptors 18
Pieridae 10, 11, 17, 30, 58, 113–131
  eggs and caterpillars 20, 21, 114
Pierinae 113, 126–129
  small tribes 130–131
Pierini 126–129
Pieris rapae 56, 128
Pinacopteryx eriphia 131
Pinacopteryx eriphia tritogenia 112, 113
Pirate 202, 203
Piruna aea 105
plant families, for caterpillars 21, 26–27, 31, 61, 80, 89
Plastingia naga 109
Platylesches neba 108
Polka Dot 181
Polygonia 202
Polyommatini 144, 160–161
Polyommatus bellargus 160
Popinjay 187
Poplar Admiral 27
Poritia hewitsoni 153
Poritiinae 144, 152–153

Poritiini 152–153
Praepapilio spp. 81
Praepapilioninae 63, 81
Precis octavia 201, 202
Prepona (Agrias) 6, 7, 44, 210–211
Prepona (Agrias) claudina 210, 211
Preponini 210
proboscis (tongue) 14, 34, 110
Prodryas persephone 41
prolegs 21
Prothoe franck 211
Prothoini 210
prothorax 14
Protogoniomorpha parhassus 201, 202, 203
Pseudacraea 177
Pseudacraea lucretia 177
Pseudacraeini 177
Pseudergolinae 186–187
Pseudergolis 187
Pseudergolis wedah 187
Pseudohaetera hypaesia 213, 214
Pseudoneptini 177
Pseudopontia 124–125
Pseudopontia australis 125
Pseudopontia paradoxa 125
Pseudopontiinae 113, 124–125
Pterourus 80
Pterourus (Papilio) homerus 36, 48
publishing, journal 53
puddling 10, 23–24, 75, 114, 123, 190, 207, 210
Punchinello 132, 133
pupa(e) 19, 22–23
  camouflage 23, 33
  development 22–23, 47
  Hesperiidae 89, 90
  Papilionidae 64
  Pieridae 114
Purple and Gold Flitter 110
Purple Emperor 189, 190
Purple Sapphires 154
Pyrginae 100–101
Pyrgus 100
Pyrrhiades lucagus 91
Pyrrhopyge papius 99
Pyrrhopyginae 98–99
Pyrrhopygini 98–99
Pythonides jovianus 88

Q

Quadrus cerialis 101
Queen Alexandra's Birdwing 30, 45, 48, 58, 77

R

Red-banded Hairstreak 35, 159
Red Lacewing 184
Red Pierrot 161
Red Underwing Skipper 101
Regent Skipper 92, 93
research, butterfly 46–49

ACKNOWLEDGMENTS

The authors would like to thank David Price-Goodfellow, Joanna Bentley, Ginny Zeal, and Susannah Jayes at Bright Press, as well as Frances Cooper and a reviewer for their support through this project. Many thanks to all those individuals and institutions that provided images for use in the book.

Blanca would like to thank the many entomologists and naturalists capturing their ideas and sharing their knowledge in the hundreds of papers, images, data, and books consulted when writing this book. Thanks to the Natural History Museum (NHM), London, for preserving the memory of our planet

through its amazing scientific collections. Thanks to NHM colleagues Gavin Broad, Jo Wilbraham, Richard O'Brien, Tim Littlewood, and Colin Ziegler.

My deepest gratitude goes to my husband, son, and parents for their support and encouragement to make this book possible. And to the butterflies and the jungles in Colombia for giving me a life-long curiosity and passion.

To Berry and Coco.

PICTURE CREDITS

The publisher would like to thank the following for permission to reproduce copyright material:

T = Top; B = Bottom; L = Left; R = Right; C = Center.
Aga Pierwola, Department of Invertebrate Zoology, American Museum of Natural History: 53CL, 53BL. **Alamy Stock Photo** Mike Denton: 6; SeaTops: 10TL; Kelley Miller: 24TL; Oliver Thompson-Holmes: 33BCR; Wirestock, Inc: 34TL; INTERFOTO: 34TR; Glyn Thomas: 36CR; Alain Guilleux: 54; blickwinkel: 60, 68; Danita Delimont: 75TR; DEEPU SG: 77; Survivalphotos: 78; All Canada Photos: 79T; imageBROKER.com: 101BL; Bryan Reynolds: 105TR; Rick & Nora Bowers: 105CL; FLPA: 107; Colin Marshall: 109B; AGAMI Photo Agency: 112; imageBROKER.com: 142; Minden Pictures: 148; Rick & Nora Bowers: 149T; Premaphotos: 150; Bryan Reynolds: 159BL; Imagery India: 161TL; Bryan Reynolds: 166; Natalia Kuzmina: 167T; blickwinkel: 168; Bryan Reynolds: 174; Ephotocorp: 175TR; ZUMA Press, Inc.: 181BL; Ephotocorp: 187BR; Daniel Borzynski: 190; Arto Hakola: 192; Minden Pictures: 194T; Steve Holroyd: 194CL; David Smith: 195T; CFC Collection: 214TR; praveen g nair: 218; Morley Read: 223TL; Malcolm Schuyl: 223CL; Danita Delimont: 225; Gabbro: 227TL; Robert Kennett: 228; Lars S. Madsen: 229CR; Gabbro: 230; BIOSPHOTO. 231TL; Louise Heusenkveld: 231CL. **Andrew Neild**: 9, 13T (*Heliconius pardalinus*), 13CT (*Heliconius elevatus*), 99TR, 134, 135, 141, 158, 170CR, 210, 217TR, 227TR. **Antonio Giudici**: 73. **Antonio Robles**: 155TR. **Armin Dett**: 103. **Bill Berthet**: 31BL. **Blanca Huertas**: 59C, 61 ,124, 224. **Bridgeman Images** Look and Learn: 44. **Carnegie Museum of Natural History, Invertebrate Zoology Archives**: 45TL. **Cheng-Chia Tsai**: 15, 48. Images courtesy of **Daniel Jaramillo F. @dajafer**: 82B, 88CL, 182. **David Geale**: 94, 118, 227B. **David Price-Goodfellow**: 203TR, 216. **Don R Simons**: 29TL. **Dreamstime** EPhotocorp: 33TCR; Roshan Thapa: 33BR; Nuwat Phansuwan: 88CR; Matee Nuserm: 90; Xiaoxia Jia: 95TL; Thawats: 109T; Shubhrojyoti Datta: 147B; Rajashrijoshi: 157T; EPhotocorp: 186. **Ernst Mayr Library and Archives of the Museum of Comparative Zoology, Harvard University**: 45TR. **Frank Model**: 93. **Fredy Montero Abril**: 165, 219BL, 220. **Galeria Amaya** (www.galeriaamaya.com): 56. **Geoff Gallice** (Flickr): 119. **Getty Images** Photo by DANIEL MUNOZ/AFP: 27TR; Daily Express Archive: 158. **iNaturalist** desertnaturalist: 7; German Leonel Sarmiento Cruz: 13CL; Cheongwei Gan: 65, 72; Rob Foster: 76; Cheongwei Gan: 123TL, 132; desertnaturalist: 136BCR, 140CR; renjus box: 147TL; dineshphotography7797: 151; Michael J Papay: 159TR; Christoph Moning: 160CR; Pablo Bombin: 189BR; Cheongwei Gan: 199TR. **iStockphoto** mtreasure: 27TL; sirichai_raksue: 179CR. **Japan Butterfly Conservation Society**: 39. **Jérôme Albre**: 67TL. **Jocelyn Wang, Trey Scott, Naomi Pierce**: 35TL. **Keith Willmott**: 51. **Lawrence Reeves**: 35BL. **Luc Legal**: 67TR. **mauritshuis.nl**: 55BL. **Minneapolis Institute of Art**: 42BL. **Museum of Comparative Zoology, Harvard University** © President and Fellows of Harvard College: 41TL. **National Academy of Sciences, USA** © 2004: 46. **Natural History Museum, London**: 31TL, 40, 58, 59TL, 82B, 84, 102. **Nature Picture Library** Pete Oxford: 23TL; Iphiclides podalirius: 23TR; Thomas Marent: 34CL; Jussi Murtossaari: 86; John Abbott: 95CL; Paul Harcourt Davies: 104CL; Steve Knell: 104CR; Suzi Esterhas: 114; Claudio Contreras: 128; Silvia Reiche: 130; Hanne & Jens Eriksen: 131TR; Thomas Marent: 131CR; Emanuele Biggi: 137; Paul Bertner: 138CL; Thomas Marent: 138CR, 139; David Tipling: 161TR; Lynn M Stone: 162; Thomas Marent: 164; Michael & Patricia Fogden: 170CL; Aflo: 172; Doc White: 173; Nature Production: 177C; Stephen Dalton: 180; Ingo Arndt: 183CR; Alex Hyde: 189T; Thomas Marent: 197CL; Luiz Claudio Marigo: 197BR; PREMAPHOTOS: 201CR; Edwin Giesbers:

202BL; Guy Edwardes: 204; Alex Hyde: 205C; Robert Thompson: 208; Luiz Claudio Marigo: 211B; Piotr Naskrecki: 221BR; Glass and Nature: 222. **Pablo Martinez-Darve Sanz**: 100, 120, 140TR, 183TR, 183BR, 199CR, 215T, 215CR. **Peter Bygate** (www.lepidigi.net): 125, 185BR. **Public Domain**: 32, 43TR. **Quindio Botanical Garden, Colombia**: 57. **Richard Sehnal, ČZU – Forbio**: 66. **Sebastián Padrón**: 16, 17, 24TR, 34CL. **Shinichi Nakahara**: 18, 47. **Shutterstock** David Havel: 2; HHelene: 10BL; Super Prin: 13C; Pacotoscano: 25; Bildagentur Zoonar GmbH: 26; SAI ADA: 33TR; Boris15: 36CL; svic: 36BL; Lucian Coman: 62; Anson 0618: 64; Reflex Nature: 69TR; Matej Ziak: 69C; featthercollector: 71; Ondrej Prosicky: 74; Thanit Werrawan: 75CL; Wirestock Creators: 79C; James Hou: 88TR; Luis_zapata: 95CR; Steve Byland: 98; Kobus Peche: 105CR; Flecksy: 116; Alec Issingonis: 117; MG East: 121; Stephane Bidouze: 122; addi2020: 123B; Stephane Bidouze: 127; Worraket: 145B; Shubhrojyoti: 146; Martin Fowler: 155C; Wirestock Creators: 157CR; samray: 159BR; Dina Rogatnykh: 167BR; Eko Budo Utomo: 169; Stephane Bidouze: 178; tharamust: 179BL; Wagner Campelo: 185CL; Dark Egg: 187CR; James Hou: 191; haraldmuc: 193; khlongwangchao: 195BR; Nyanews: 200; Michael Overkamp: 201TR; JJ van Ginkel: 202BR; Stephane Bidouze: 203BL; Jukka Jantunen: 203BR; Kevin Collison: 205TR; Mark Frankcombe: 209; Saharath Kachacupt: 211CR; David Havel: 212; Martin Pelanek: 214CL; Ondrej Prosicky: 221CR; Alejandro Alves Mendes: 226; vancahi: 229T; Super Prin: 229CL; Matee Nuserm: 231BR. **Silvana Bossa and Julio Villadiego/ Courtesy of Gabo Foundation**: 55T. **US National Park Service**: 37. **Vitaly Charny**: 52. **Wikimedia Commons** Quarti: 5; Frontispiece from *The Naturalist on the River Amazons*, H.W. Bates, 1863: 8; Alexey Yakovlev: 29BR; Peelldeen: 30; Biodiversity Heritage Library: 41BL; staedelmuseum.de: 42BR; Nationalmuseum, Stockholm: 43TL; Lowe, Tristan, University of Manchester, Garwood, Russell J., University of Manchester, Simonsen, Thomas J., Natural History Museum, Bradley, Robert S., University of Manchester, Withers, Philip J., University of Manchester: 49; MaheshBaruahwildlife: 70T; Charles J. Sharp (sharpphotography.co.uk): 70BL; JJ Harrison (www.jjharrison.com.au): 75CR; Vijayanrajapuram: 80; Pavel Kirillov: 82T; Tommy Andriollo: 85; Cláudio Dias Timm: 88TL; Peelldeen: 89; Charles J. Sharp (sharpphotography.co.uk): 91T; Vkchandrsekharanlic: 91CL; Dr Raju Kasambe: 91CR; Greg Tashney: 92; BrijeshPookkottur: 94BL; Charles J. Sharp (sharpphotography.co.uk): 96CR; ManaskaMukhopadhyay: 97; desertnaturalist: 99C; Judy Gallagher: 101TL; Charles J. Sharp (sharpphotography.co.uk): 101TR; Alex Popovikin: 101CL; Graham Winterflood: 106; Benard DUPONT: 108; Arky1993: 110CL; Orchi: 110BL; Alan Schmierer: 111; Firex AK: 115; Charles J. Sharp (sharpphotography.co.uk): 126, 129; Gilles San Martin: 136TCR; Christian Pirkl: 145T; Atanu Bose Photography: 149BR; Charles J. Sharp (sharpphotography. co.uk): 152CL; JMK: 152CR; Sourabh.biswas003: 153TR; Charles J. Sharp (sharpphotography.co.uk): 153CL, 153CR; Sandipoutsider: 154, 156; Charles J. Sharp (sharpphotography.co.uk): 159CR; Diliff: 160T; David Tiller: 171; Charles J. Sharp (sharpphotography.co.uk): 175C, 177T; Jusy Gallagher: 181CL; Atanu Bose Photography: 184; Peellden: 188; Charles J. Sharp (sharpphotography. co.uk): 196; Sourabh.biswas003: 198; Peelldeen: 206; Antanu Bose Photography: 207; Charles J. Sharp (sharpphotography.co.uk): 213TR; © 2017 Jee & Rani Nature Photography: 213CR, 214TL; Andrew Allen: 217C; Olei: 219TL. **Yeison Vega**: 13B.

All reasonable efforts have been made to trace copyright holders and to obtain their permission for the use of copyright material. The publisher apologizes for any errors or omissions and will gratefully incorporate any corrections in future reprints if notified.